D1334541

CROP NUTRITION AND FERTILISER USE

CROP NUTRITION AND FERTILISER USE

by
JOHN ARCHER

FARMING PRESS LTD
Wharfedale Road, Ipswich, Suffolk

First published 1985
Second edition 1988

Copyright © Farming Press Ltd, 1985, 1988

Distributed in North America
by Diamond Farm Enterprises,
Box 537, Alexandria Bay, NY 13607, USA

Archer, John
Crop Nutrition and fertiliser use.— 2nd ed.
1. Field crops   2. Plants——Nutrition
I. Title
633   SB185
ISBN 0-85236-75-0

ACKNOWLEDGEMENTS

*Extracts from HMSO and MAFF publications are
reproduced with the permission of the Controller of
Her Majesty's Stationery Office.*

Phototypeset by Galleon Photosetting, Ipswich
Printed and bound in Great Britain by Page Bros (Norwich) Ltd.

CONTENTS

ILLUSTRATIONS

COLOUR PLATES

PLATES

FIGURES

PREFACE

IN RECENT years an ever increasing range of fertiliser products, services and recommendations have been promoted in the UK. An understanding of nutrient behaviour in soils and uptake by crops is essential in ensuring that a fertiliser recommendation will provide both an economic crop response and a longer term maintenance of soil fertility. In *Crop Nutrition and Fertiliser Use* I have attempted to provide the background understanding necessary for making decisions on fertiliser use for the crops and soils of the UK.

Agriculture and horticulture are fortunate in this country in having a substantial experimental basis for fertiliser recommendations for a very wide range of crops. Much of this work is carried out by my colleagues in the Agricultural Development and Advisory Service (ADAS). This work is complemented by Rothamsted Experimental Station, which has provided a world lead in fertiliser research for over a century.

In preparing this book, I have been helped by numerous friends and colleagues in ADAS, Agriculture and Food Research Council (AFRC) Institutes and in the fertiliser industry. Their considerable assistance in providing material is acknowledged with grateful thanks. My special thanks are extended to Bryan Davies for his encouragement, to Roger Unwin for his critical comment on the text and to my wife Sandra for her considerable help and enthusiastic support.

June 1984   *John Archer*

Chapter 1

INTRODUCTION

THE AIM of this book is to provide a comprehensive coverage of the principles of crop nutrition and fertiliser recommendations for field crops grown in England and Wales. Before considering the individual nutrients and crops, this chapter is concerned with both an historical perspective and the future use of fertilisers and their possible limitations for economic or environmental reasons.

HISTORY OF CROP NUTRITION

An understanding of the role of the major nutrients in plant nutrition had been achieved by 1840. Work in Germany by Liebig and at Rothamsted by Lawes and Gilbert had shown the benefits to crop growth when the major plant nutrients were added to the soil in the form of fertilisers. Nitrogen, phosphorus, sulphur, potassium, calcium and magnesium were used as the main additives in water culture experiments of plant growth carried out in the mid 1800s.

Acceptance of the need for the various micronutrients or trace elements has been established since this initial work. Iron was confirmed as a plant nutrient in the mid 1800s when it was shown that it could cure what is now called iron-induced chlorosis. Manganese was found to be essential for fungal growth as early as 1863; however, it was not until the 1920s that manganese was shown to cure grey speck of oats and then, in 1941, marsh spot in peas. Zinc was seen to be essential for the growth of maize during the early 1900s and the needs for boron and copper were established during the 1920s and 1930s. The importance of molybdenum was only fully appreciated in the 1940s following work at Long Ashton. The most recent addition to this list is chlorine which was only shown to be essential in 1954. The reason for such a long delay was probably the lack of field deficiency problems. Cobalt was also discovered to be important for legumes at much the same time. Numerous other elements have been, and still are being, studied but the likelihood of further additions to the list of essential elements declines with time. It is improbable that any further additions will now be proved as most possible contenders have been thoroughly researched and ruled out.

HISTORY OF UK FERTILISERS

The benefits of farmyard manure, chalk and marl have been appreciated, if not understood, for several hundred years. In 1843 the first use of superphosphate occurred following field experimental work and initial production by Lawes and Gilbert at Rothamsted. This work showed the benefits of phosphate for a range of crops. It should be remembered that at this time most soils were deficient in phosphate. Few UK soils are naturally well supplied with phosphate; the only fields not deficient would be those which had received appreciable quantities of farmyard manure. Other organic sources of phosphorus were animal and fish processing wastes which were sold as manures and imported guano, which is composed mainly of bird droppings. The only significant inorganic source of nitrogen was the imported, naturally occurring sodium nitrate (Chilean nitrate of soda).

The modern UK fertiliser industry started in the 1850s with the manufacture of superphosphate. The first sources of phosphate were crushed bones and phosphatic nodules called coprolites which occur in the Greensand and associated strata. The main area of quarrying was in south Cambridgeshire and the adjoining counties. The coprolite bed occurring between the Chalk and the Gault Clay is unique to the south Cambridgeshire area. The heyday of superphosphate production from coprolites was reached in the mid 1870s. By the mid 1880s, imported phosphate ores were processed by the coprolite plants and the decline of quarrying set in. The coprolite rush was over by 1900. Processing the coprolites into superphosphate consisted of the simple but laborious process of washing the nodules free from clay particles followed by grinding and treatment with sulphuric acid. The imported ore was much easier to process since no washing was necessary.

Another important source of phosphate was basic slag. A by-product of steel making in England and Wales, this was widely used on grassland until the late 1970s when home-produced iron ore was superseded by high-grade imported ore which does not produce a worthwhile phosphate fertiliser by-product.

During the second half of the nineteenth century, ammonium sulphate became available in increasing quantities as a by-product of town gas production. This was the main source of inorganic nitrogen for many years, but was replaced during the early twentieth century by the direct fixation of atmospheric nitrogen to form ammonia (the Haber process). This process for ammonia production started commercially in Germany in 1913 and by 1920 was being used in both Europe and America as the major method of ammonia production for use in nitrogen fertilisers. In 1965 ICI introduced ammonium nitrate in quantity to the UK market.

By contrast, potassium fertilisers were available as natural salts and were processed as muriate of potash during the late nineteenth century from the Strassfurt mines in Germany. Today most is imported from Europe or Canada, with hopes of increased production from the Cleveland deposits of northern England.

FERTILISER USAGE

All nutrient recommendations in this book are quoted in kilograms per hectare (kg/ha) of nutrient. A frequent source of confusion is that both rate of nutrient and rate of product are commonly quoted in the same units – kg/ha. To convert kg/ha of nutrient to units/acre, the figure should be multiplied by 0.8, ie. 50 kg/ha = 40 units/acre. The number of kg of nutrient in a standard 50 kg bag of fertiliser is given by dividing the percentage of that nutrient by two, ie. a 50 kg bag of 20:10:10 NPK compound contains 10 kg N, 5 kg P_2O_5 and 5 kg K_2O or a 50 kg bag of 34 per cent N contains 17 kg of N.

> 1 kg/ha = 0.8 units/acre
> 1 unit/acre = 1.25 kg/ha.

The unit is defined as 1.12 lb or 1 per cent of 1 cwt.

Throughout the book the general usage oxide expression of nutrients is used for phosphorus (P_2O_5) and potassium (K_2O) and less frequently for calcium (CaO) and magnesium (MgO). These means of expression are required by the Fertiliser Regulations but are historical and have no scientific basis to justify their use.

> $P_2O_5 = P \times 2.3$ $P = P_2O_5 \times 0.4$
> $K_2O = K \times 1.2$ $K = K_2O \times 0.8$
> $CaO = Ca \times 1.4$ $Ca = CaO \times 0.7$
> $MgO = Mg \times 1.7$ $Mg = MgO \times 0.6$

The figures in Table 1.1 show the changes in the patterns of nitrogen (N), phosphate (P_2O_5) and potash (K_2O) use over the last 140 years. The greatest increases in nitrogen use took place during World War II and again during the 1970s. Phosphate usage doubled during World War II but has remained fairly constant since then. Very little potash was used before World War I. During and after World War II, the use of potash increased steadily but has been on a plateau since the 1960s.

Most fertilisers were applied as straights until after World War II

Table 1.1. UK fertiliser use (1000s tonnes of nutrient)

	Nitrogen (N)	Phosphorus (P_2O_5)	Potassium (K_2O)
1837	—	15	—
1845	33	46	—
1874	34	90	3
1896	33	122	5
1913	29	180	23
1929	48	198	53
1939	60	170	75
1945	172	346	115
1949	185	419	196
1958	315	386	348
1965	565	479	425
1969	790	476	458
1973	932	474	428
1977	1,092	406	409
1980	1,268	427	447
1984	1,588	488	559
1986	1,572	434	510

Source: Fertiliser Manufacturers' Association.

when granular compounds became available. Before this time, mixtures of ammonium sulphate, superphosphate and muriate of potash were made up by local merchants to the farmer's requirements. It was important that the mixtures were spread immediately. If rain prevented spreading, these mixtures of crystalline substances would absorb moisture overnight and would subsequently set in a solid heap, needing considerable energy with a sledge hammer before it could be used.

SURVEY OF FERTILISER PRACTICE

Information on fertiliser practice in England and Wales has been collected from farm surveys since the 1940s. The survey is now carried out annually under the joint sponsorship of the Fertiliser Manufacturers' Association/Agricultural Development and Advisory Service/ Rothamsted Experimental Station. Table 1.2 shows how fertiliser practice on winter wheat, maincrop potatoes and grass leys has changed since 1943/5. While the figures are not strictly comparable due to variation in the way in which the surveyed fields were selected and grouped, overall trends can be deduced.

There has been a continuing increase in nitrogen use for winter wheat but little change in phosphate or potash usage since 1960. The pattern for long-term leys has been similar. Potatoes have traditionally

Table 1.2. Average overall NPK usage in England and Wales

	N kg/ha	P_2O_5 kg/ha	K_2O kg/ha
Winter wheat			
1943/5	19	30	3
1950/2	33	28	15
1957	51	30	33
1962	74	36	43
1966	90	44	44
1974	91	46	38
1978	125	44	37
1981	162	49	42
1986	186	56	52
Maincrop potatoes			
1943/5	79	91	100
1950/2	117	124	165
1957	124	125	199
1962	160	148	245
1966	161	173	241
1974	175	183	237
1978	186	193	252
1981	194	192	259
1986	195	203	264
2–7-year leys			
1943/5	4	11	0
1950/2	16	35	15
1957	26	34	21
1962	54	43	33
1966	67	46	33
1974	133	36	29
1978	161	34	34
1981	172	32	39
1986	189	29	47

Source: Survey of Fertiliser Practice.

received large amounts of inorganic phosphate and potash, often in addition to farmyard manure. The table shows the gradual increase in NPK usage on maincrop potatoes over the last forty years. Data from the last few years reveal that liquid fertilisers comprise about 8 per cent of the England and Wales market. In 1978, combine drilling was still very popular with 60 per cent of the winter wheat and two-thirds of the spring barley being combine drilled.

Data from the survey indicate that overall usage comprises about two-thirds compounds and one-third straights. Nitrogen as straights is increasing while phosphate and potash are decreasing. The 1980 figures for straights as a percentage of the total nutrient market are N at 56 per cent, P_2O_5 at 13 per cent and K_2O at 2 per cent.

NUTRIENT PRICES

While there is considerable variation in the cost of nutrients to the farmer in any one year, Table 1.3 shows the trends in prices since 1931. It emphasises the stable price situation until World War II. The cost of

Table 1.3. Cost of nutrients

	N p/kg	P_2O_5 p/kg	K_2O p/kg	Cost kg N/Value kg wheat
1931	4.5	2	1.5	7.5
1940	4.0	2	1.5	4.0
1950	6	4	3	2.3
1952	8	7	3	2.7
1957	10	8	3	4.8
1963	10	8	3	2.0
1973	10	10	6	3.3
1974	16	22	10	2.9
1975	16	28	10	2.7
1977	19	25	10	2.2
1980	29	34	15	2.9
1982	35	28	16	3.2
1986	27	26	17	2.8

nitrogen and phosphorus increased in the early 1950s along with a general increase in prices, but potash did not change. The price of nitrogen is closely related to oil prices explaining the dramatic increase over the last ten years. The sharp increase in the price of phosphate compared to the other nutrients during the early 1970s followed a general increase in price by the main exporters of rock phosphate. Potash has increased at less than the rate of inflation in recent years due to excess world production.

The costs in Table 1.3 are gross costs. From World War II until 1974, inorganic nitrogen and phosphate fertilisers received a direct government subsidy but no subsidy was paid on potash fertilisers.

VALUATION OF FERTILISER RESIDUES

The Agricultural Holdings Act 1948 gave an outgoing tenant the statutory right to compensation for improvements made to a farm. Recent applications of lime, phosphate, potash and farmyard manure were eligible. The most recent updating of the amounts of compensation were made in 1983 (Statutory Instrument No 1475).

The scheme for the valuation of recent applications appears in Table 1.4. The lime values are based on the cost of lime at the time of application. The figures for phosphate residues are calculated assuming two-thirds left after one crop, one-third after two crops and one-sixth after three crops. Nothing is paid after three years. The comparable figures for potash are one-half, one-quarter and nil after three growing seasons. The only compensation for nitrogen is on organic materials, excluding blood and urea. These are assumed to have one-half value after one year and one-quarter after two years. Composite figures are used for valuation of farmyard manures, calculated on assumed nutrient contents as discussed in Chapter 11. Compensation is also allowed for the unexhausted manurial value of animal feeding stuffs used over the last two seasons on the farm.

The whole rationale of the compensation scheme is one of land improvement rather than of maintaining soil pH and nutrient levels. The latter would seem to be a more appropriate concept in the 1980s.

ORGANIC FARMING

A full discussion of organic farming is beyond the scope of this book. In most interpretations of organic farming, inorganic sources of nutrients are not used. Reliance is put on recycling nutrients and using organic sources when additions are needed. As far as crop nutrition is concerned, there is no reason why yield levels on medium or heavy

Table 1.4. **Valuation of lime, phosphate and potash residues**

Depreciation of value of lime

Mean annual excess winter rainfall (mm)	Where the land comprises permanent pasture or long-term leys with more than 250 kg per hectare of nitrogen applied annually, or arable or mixed ley and arable land	Where the land comprises permanent pasture or long-term leys with up to 250 kg per hectare of nitrogen applied annually
	Value of lime to be depreciated over:	Value of lime to be depreciated over:
Less than 250	8 years	9 years
250 to 500	6 years	7 years
More than 500	4 years	5 years

Unit value of phosphoric acid (as P_2O_5) in one per cent of a tonne of fertiliser

Nature of fertiliser	After growing season		
	One	Two	Three
	p	p	p
1. Organic forms and inorganic forms (including basic slag) but excluding rock phosphates and calcined aluminium calcium phosphate	158	79	39
2. Soft ground rock phosphates applied in:			
(a) areas with a mean annual excess winter rainfall of 450 mm or more	158	79	39
(b) areas with a mean annual excess winter rainfall of less than 450 mm			
(1) Permanent grassland	158	79	39
(2) Other crops	nil	nil	nil
3. Other ground rock phosphates applied in:			
(a) areas with a mean annual excess winter rainfall of 450 mm or more	39	39	39
(b) areas with a mean annual excess winter rainfall of less than 450 mm	nil	nil	nil
4. Calcined aluminium calcium phosphate	The value, if any, shall be such as may be determined in accordance with scientific evidence		

Unit value of potash (K_2O) in one per cent of a tonne of fertiliser

Type of crops to which fertiliser is applied	After growing season		
	One	Two	Three
	p	p	p
1. Applied to arable crops (except forage crops) and all root crops where tops are left on the land, except potatoes (see also 4 below)	92	46	nil
2. Applied to leys, permanent grassland or forage crops which are grazed or the product cut and fed on the holding	92	46	nil
3. Applied to leys and permanent grassland the product of which is cut and removed from the holding	nil	nil	nil
4. Applied to roots (including potatoes) and forage crops which are removed from the holding	nil	nil	nil

Source: Landlord and Tenant Legislation, 1983.

soils should not be similar to those achieved with inorganic nutrient sources. Even on sandy soils, nitrogen and potash from animal manures can achieve good yields.

The principles of crop nutrition remain the same, whether nutrients are provided in inorganic or organic form. The main need is to supply adequate nitrogen. By using legumes and animal manures, most organic farming systems have no problem in providing enough nitrogen for high-yielding crops. Any organic farm is likely to need animal manures either bought-in or produced on the farm. Legume-only rotations are unlikely to be practicable.

As long as the soils are not deficient in phosphorus, this nutrient is unlikely to be a limitation. In most cases phosphorus in bought-in animal feeds will at least balance that removed from the farm in crop and animal products. Bought-in animal manures or sewage sludge offer other possibilities of maintaining soil phosphorus levels.

On clay soils, potassium will generally be supplied by the natural weathering of clay minerals. Problems can easily arise on sandy soils if insufficient potassium is applied. As long as efficient handling and use of animal manures is achieved without loss of urine, which contains appreciable potassium, this problem will be minimised. However, low soil potassium levels are commonly the main limitation when attempting organic farming on light soils.

ENVIRONMENTAL PROBLEMS

Whilst wind and water erosion lead to small amounts of soil reaching streams and rivers, it is the more soluble nutrients in soils that are of major concern in causing environmental problems. Algal blooms or excessive growth of aquatic weeds occur in lakes and streams under some conditions. For this to happen an adequate concentration of nitrogen and phosphorus in the water is necessary. Research has shown that nitrogen levels are invariably above this threshold, but phosphorus is often the limiting nutrient. While some phosphorus finds its way into water by direct soil movement, the major sources of water-soluble phosphorus in streams come from sewage effluent or direct contamination with faeces either human or animal. Inorganic fertiliser use is not a major factor in this form of pollution.

The nitrate content of drinking water is a topic of general concern. Where water is abstracted from streams and rivers, nitrate is contributed by land drainage water. Aquifer water nitrate levels are influenced by the agricultural practice above the aquifer. Nitrate levels in aquifer waters have generally increased in the UK over the last thirty years. The EEC limit for drinking waters is 50 milligrams per litre (mg/l) of nitrate, but some boreholes are above this and must be blended before distribution.

Much research effort has gone into trying to elucidate the various contributing factors to nitrate levels in stream and aquifer waters. Large amounts of nitrate are lost following ploughing up of grassland with a consequent oxidation of much organic nitrogen over several years. Legumes and animal manures increase the soil nitrogen supply and can result in increased nitrate leaching. Nitrogen fertiliser use is also important, but the extent depends on the circumstances of weather, cropping and soil type. Well-timed nitrogen applications reduce the amount of nitrogen leached as nitrate under arable cropping. Losses are generally less under grass and also lower on medium and heavy soils than on sands or shallow soils over rock. Factors affecting the nitrate content of water from field drainage systems are better understood than the factors involved in contributions to ground waters.

The Department of the Environment published a report 'Nitrate in Water' Pollution Paper No 26 in 1986. It highlighted the main areas with current high nitrate water supply problems as East Anglia, Lincolnshire, Nottinghamshire and Staffordshire. As more water supplies increase in nitrate concentration, the possibility of blending high and low nitrate sources to keep below the nitrate threshold level becomes more difficult. Other options are denitrification plants to remove nitrate from individual sources and nitrate protection zones which would limit nitrogen fertiliser use over part or all of the catchment area of a particular borehole.

The report lists the following measures which could be adopted generally to reduce leaching with minimum effects on farm profitability:

- no application of nitrogen fertiliser from mid-September to mid-February.
- planting of autumn-sown crops in preference to spring crops.
- planting of winter cereals as early as possible in the autumn.
- maintenance of crop cover in autumn and winter.
- careful field-by-field assessment of the quantity and timing of nitrogen fertiliser application (both inorganic and organic).

The report suggests that localised problems might be dealt with by the following measures to protect particular sources:

- more attention given to explaining to farmers what practical steps they can take to reduce nitrate leaching.
- alternative land uses, e.g. grassland, forestry, recreation.
- leaving grassland unploughed.
- limitation on fertiliser application rates.

The third aspect of current concern is the nitrate content of food crops. Leafy crops will contain some nitrate whether grown using inorganic or organic nitrogen sources. How much depends partly on weather conditions and, more importantly, on the stage of maturity. Young plants tend to accumulate nitrate prior to enzymic conversion to ammonium–N and subsequent incorporation into proteins. Spinach is often high in nitrate. A maximum nitrate content is applied to decide whether it is suitable for use in baby foods. Mature roots, fruits and seeds contain very little nitrate.

FUTURE USE

To date, the use of fertilisers in the UK has been dictated almost entirely by economic factors. Around 1970 it was generally assumed that any major increase in fertiliser use would be nitrogen on grass. Nitrogen use has increased considerably since this time but economic circumstances have directed much of it to winter cereals and oilseed rape as well as grass.

Overall nitrogen use in the UK has reached a plateau in the last few years. This has resulted in the first small fall in annual nitrogen fertiliser sales. Rates of use on arable crops are unlikely to increase further. Any increase in the area of peas and beans will reduce overall nitrogen fertiliser use. While the scope for increased use on grass remains, the introduction of quotas on milk production means that few dairy farms are likely to increase their nitrogen use over the next few years. Current EEC food surpluses could result in a considerable reduction in UK fertiliser use if substantial areas of land are taken out of food crop production.

Major changes in nitrogen use on livestock farms may be forced by environmental pressures – voluntary or statutory – to use animal manures more efficiently in order to reduce losses of nitrogen. Initially any controls on nitrogen usage are likely to be implemented through voluntary codes of practice, particularly concerning the timing of inorganic and organic nitrogen applications. The necessary legislation for this is already on the statute book in the 1974 Control of Pollution Act.

Phosphate use is unlikely to show a general increase. Its use in grassland areas may increase slightly, but use in arable areas is likely to decline where soil levels are high and increase slightly in other arable areas as yields increase. Similarly potash use is only likely to increase if more is used on grassland. Small changes in arable use are unlikely to have a significant effect on overall usage.

Chapter 2

PLANT NUTRIENT UPTAKE

PLANTS NEED a number of essential elements to enable them to grow and reproduce. The quantities of these elements required allow an arbitrary distinction to be made between the major nutrients taken up and needed in much larger quantities than the minor nutrients, sometimes called micronutrients or trace elements. All are equally essential to plant metabolism. Without any one of them, plant growth cannot take place.

ESSENTIAL NUTRIENTS

In addition to carbon, hydrogen and oxygen taken up from air and water, higher plants need an adequate supply of the nutrients listed in Table 2.1 to grow and reproduce. These are the essential elements. Some plants also benefit from other non-essential elements such as sodium. These are not the only elements taken up by plants; a very wide range will be taken up if found in the soil in which the plant is growing. Where the crop is used for animal feed, particularly grass, the uptake of additional elements as indicated in Table 2.1 may be very important for animal production.

Some elements are toxic to plants and if the soil contains high levels of one or several of these elements, plant growth may be adversely affected. Others may adversely affect the animal feeding on the crop. The plant essential nutrients are discussed individually in succeeding chapters. Aspects of those elements required specifically by animals are contained in Chapter 15. Elements that are toxic to plants are discussed in the sewage sludge section of Chapter 11 on organic manures.

In Table 2.1 the main ionic forms in which the elements are taken up from the soil are shown. Table 2.2 gives the quantities likely to be removed in the above ground parts of the plant at harvest. There is

Table 2.1. Elements essential for plants and animals and form in which absorbed by roots

Element		Essential for plants	Essential for animals
Nitrogen	NH_4^+, NO_3^-	yes	yes
Phosphorus	HPO_4^{2-}, $H_2PO_4^-$	yes	yes
Potassium	K^+	yes	yes
Calcium	Ca^{2+}	yes	yes
Sulphur	SO_4^{2-}	yes	yes
Magnesium	Mg^{2+}	yes	yes
Sodium	Na^+	no	yes
Iron	Fe^{2+}, chelates	yes	yes
Manganese	Mn^{2+}	yes	yes
Boron	$B(OH)_3$	yes	no
Copper	Cu^{2+}	yes	yes
Zinc	Zn^{2+}	yes	yes
Molybdenum	MoO_4^{2-}	yes	yes
Chlorine	Cl^-	yes	yes
Iodine	I^-	no	yes
Selenium	SeO_4^{2-}	no	yes
Cobalt	Co^{2+}	legumes only	yes

Table 2.2. Amounts of major nutrients removed in crops (kg/t fresh material)

	Percentage DM at harvest	N	P	K	Ca	Mg	S
Cereal – grain	85	17.0	3.4	4.7	0.5	1.3	1.3
– straw	85	6.0	0.7	6.8	3.0	0.8	0.9
Sugar beet – roots	22	1.7	0.3	1.8	0.6	0.2	0.4
– tops	16	3.2	0.5	4.8	1.4	0.8	0.6
Potato – tubers	22	3.0	0.4	4.8	0.2	0.2	0.3
Oilseed rape	92	33.0	7.0	9.0	3.7	2.3	9.0
Grass – silage	20	6.4	0.6	4.0	1.2	0.3	0.3
– hay	85	14.0	2.6	15.0	3.4	1.0	1.0
Kale	15	3.6	0.5	4.2	3.0	0.3	0.9

considerable variability in some of these values depending particularly on soil nutrient supply and whether a particular species can control the uptake of a nutrient. Some nutrients, notably potassium, may be taken up in luxury amounts if the soil supply allows.

OTHER ELEMENTS

Silicon
Silicon is contained in huge quantities in soils, both as quartz and in clay minerals. There is not much available for plant uptake but some soluble silicon occurs in most soils. The quantity taken up varies with the species.

Silicon is taken up in considerable quantities by cereals and grasses but there is no agreement on whether it is an essential element. It is concentrated in the straw of these crops and seems to have a role in the structure of the plant. While experiments on rice have sometimes shown a yield benefit from silicon application, attempts to reduce lodging susceptibility of cereals in this country have shown no positive benefits. Even if it were shown to be essential, most UK soils are likely to contain adequate silicon.

Cobalt
Whilst not considered essential for all plants, cobalt is essential for animals and is also required by nitrogen-fixing micro-organisms. This requirement is shown by both free living organisms and those involved in symbiosis with legumes. These processes are discussed in Chapter 4 on nitrogen. The level of cobalt in the soil necessary for N fixation is extremely low and deficiency in UK soils is unlikely. The amounts required by animals are considerably greater and are discussed in Chapter 15.

SOURCES OF NUTRIENTS

With the exception of some gaseous supplies of nitrogen and sulphur which may be used directly by crops and are discussed in the appropriate chapters, nutrients are taken up by plants from the soil. The majority of fertilisers are added to the soil and taken up via the roots. Foliar feeding is covered in Chapter 14. Nutrient uptake is from the soil solution. The supply of an adequate quantity of a particular nutrient for crop growth depends on both the behaviour of that nutrient in the soil and the ability of the crop root system to utilise it. Soil water supply has a strong influence on both crop growth and nutrient uptake.

Nutrients in Rainfall
Rainfall is not pure water. It contains various elements in solid form and in solution. The quantity of some of these nutrients depends on the

Table 2.3. Amounts of nutrients deposited from the atmosphere in England (wet plus dry deposition)

	kg/ha per year		g/ha per year
Nitrogen	10.0	Iron	2,000
Phosphorus	0.3	Manganese	100
Potassium	3.0	Boron	150
Calcium	10.0	Copper	100
Magnesium	5.0	Zinc	500
Sulphur	25.0	Molybdenum	10
Sodium	10.0	Selenium	3
Chlorine	20.0	Cobalt	2

Note: levels of some elements may be much higher near coasts or sources of industrial emission.

proximity of the sea and on industrial and urban emissions to the atmosphere. Appreciable quantities of several nutrients are added to UK soils in rainfall. Some idea of the quantities involved is given in Table 2.3.

SOIL NUTRIENT AVAILABILITY

Nutrients are taken up from the soil solution. The concentration of a nutrient in the soil solution depends on the quantity in the soil and its individual behaviour. Calcium and nitrate are often in relatively high concentrations, while phosphorus is invariably very low. Potassium and magnesium are generally intermediate.

To enable the crop to have an adequate nutrient supply throughout its life, it is not just dependent on soil solution concentrations at one time. The ability of the soil to replenish the supply and maintain the concentration, as ions are taken up by the crop, is equally important. If a nutrient is held very strongly in the soil it may not replenish the soil solution quickly enough. Alternatively the total amount of the nutrient in the soil may be inadequate. Fertilisers add to the overall supply of a particular nutrient in the soil and help to ensure that an adequate concentration is maintained in the soil solution.

Some nutrients, particularly nitrogen and, to a lesser extent, phosphorus and sulphur, are held and released by the soil organic matter. The availability of cations including calcium, magnesium, potassium and sodium depends mainly on the activity of clay mineral surfaces.

Cation Adsorption and Exchange

Clay mineral particles in the soil are negatively charged. These negatively charged surfaces attract and hold cations particularly Ca^{2+}, Mg^{2+}, K^+ and Na^+. When in equilibrium with the soil solution, a diffuse layer of cations occurs around the charged surfaces. The cation concentration is high at the surface and decreases with distance from the surface until a free solution, independent of the charged particle influence, is reached. Figure 2.1 shows the general pattern of cation behaviour. There is constant interchange between the free cations in solution and the cations held within the influence of the charged surface but not actually held on it. These are known as exchangeable cations. To balance the cation charge, the distribution of anions decreases from the free solution to the charged surface.

Any cation can exchange for another from the free soil solution. By cation exchange one Ca^{2+} can be replaced by two K^+. The amount of exchange that can occur depends on the strength of adsorption of a particular cation. This is governed by its own characteristics and also the type of clay mineral. When the concentration of a cation in the soil is increased by fertiliser addition, it will tend to exchange with cations already present, until a new equilibrium is formed. Sodium and other monovalent cations are easily replaced by divalent magnesium and calcium. Potassium behaves uniquely in being preferentially adsorbed. This is discussed in more detail in Chapter 6. In acid soils, hydrogen ions replace some of the other cations.

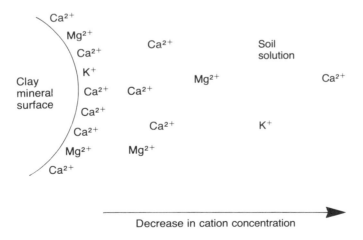

Decrease in cation concentration

Figure 2.1. Behaviour of cations in soil

Cation Exchange Capacity

The cation exchange capacity (CEC) is a measure of the total negative charge of a soil and gives a measure of its cation-retaining properties. It is measured in milliequivalents (meq) per 100 g of soil. Soils in UK vary in their CEC from under 5 meq/100 g for sands up to 30–50 meq/100 g for clays. The variation in CEC of clay soils depends mainly on the type

Table 2.4. Typical cation exchange capacities of soil constituents and soils

	meq/100 g soil
Sand and silt	3
Clay – kaolinite	5
– illite, chlorite	30
– montmorillonite	100
Organic matter	100–200
Sand	5
Light loam	10
Medium loam	20
Clay	30

of clay minerals present. Table 2.4 gives the CEC ranges for different clay minerals and organic matter. It emphasises the importance of the expanded minerals montmorillonite and illite compared to kaolinite. The balance of exchangeable cation species typically found in non-acid UK soils is of the order 70 per cent calcium, 25 per cent magnesium and 2 per cent each of sodium and potassium.

Anion Adsorption

The capacity of soils to hold anions is small compared to cation adsorption properties. Iron and aluminium oxides, clay minerals, various iron and aluminium organic complexes and calcium carbonate provide the active sites for the small anion adsorption capacity of soils. The main soil mechanism of significance is called ligand exchange with OH^- groups by which phosphate and, to a lesser extent, sulphate are preferentially adsorbed. By contrast adsorption of nitrate and chloride is extremely small. Chemical reactions particularly for phosphate also provide means of retaining anions in soil for crop use. This is detailed in Chapter 4 on phosphorus. Some nutrients particularly iron, copper, zinc and molybdenum are held in chelated forms, often as part of the soil organic matter.

LOSS OF NUTRIENTS

Various methods of nutrient removal from the soil occur. The amounts of a nutrient removed depend mainly on its individual behaviour in the soil.

The main losses are:

- removal in the crop,
- gaseous loss, mainly nitrogen,
- loss by leaching,
- soil erosion,
- irreversible soil fixation.

The importance of these losses of available plant nutrients is discussed in the appropriate nutrient chapters. All the major nutrients except phosphorus and potassium are lost in considerable quantities by leaching through the soil profile.

Leaching

For those nutrients not held strongly in the soil, leaching is an important mechanism of nutrient loss. The main leached cations are Ca^{2+}, Mg^{2+} and Na^+ with the associated anions NO_3^-, HCO_3^-, Cl^- and SO_4^{2-} (Table 2.5). Losses of all these ions can be considerable over the winter under UK conditions and fertiliser policy must take this into account.

Losses of ions occur because soil solution in equilibrium with the exchangeable cations and associated anions moves down the soil profile when rain falls on a soil already at field capacity. This rain is low in nutrients and so takes up more from the soil exchange surfaces on coming to a new equilibrium. This results in water containing a much higher nutrient concentration than rainwater being lost to the drains or beyond rooting depth and a net loss of nutrients from the soil.

The effectiveness of the leaching process depends in part on soil structure which determines how quickly rainwater comes to equilibrium

Table 2.5. Relative leaching rates of fertiliser ions in mineral soils

Cations		Anions
Sodium	*Fast*	Chloride
Calcium	↑	Nitrate
Magnesium		Sulphate
Potassium	↓	Phosphate
Ammonium	*Very slow*	

with the soil. If much of the leaching is by water movement along major fissures, as occurs in clays, nutrient loss will be much lower than from sands. If the rate of water movement is very slow, allowing the water in the fissures time to equilibrate with all the water inside the soil aggregates, this difference will not be apparent.

SOIL TYPE, TEXTURE AND STRUCTURE

Soil characteristics influence both the chemical form in which nutrients are held in the soil and the physical ability of plant root systems to explore the soil profile and extract these nutrients. Both aspects contribute to nutrient availability.

Soil texture of both topsoil and subsoil determines many of the nutrient holding properties of the soil. Figure 2.2 shows the soil texture classes as determined by the relative proportions of sand, silt and clay sized particles. Both the amount and type of the clay are important, particularly for cation holding properties of soils. Sand and silt have a considerable effect on water holding capacity but much less on nutrient holding.

The topsoil organic matter content varies from 1 to 8 per cent in mineral soils. Soils between 8 and 25 per cent organic matter are given

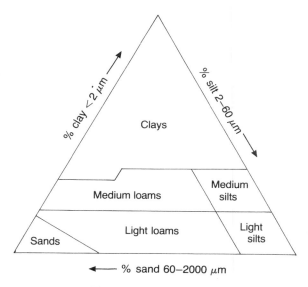

Figure 2.2. Soil textures
Source: Soil Survey of England and Wales

Plate 2.1. Nutrient enrichment and compensatory growth of part of the barley root system
Letcombe Laboratory

the prefix, organic; those above 25 per cent are called peaty. Variation in organic matter content influences water and nutrient holding capacity, ease of cultivation and root penetration. Nitrogen supply by the soil is influenced by organic matter content. Both texture and organic matter strongly influence topsoil structure and the risk of soil damage and root restriction following cultivation. Subsoil structure can have a great influence on root penetration and the ability of crops to find adequate water for growth during dry times. In particular, plough pans may severely limit root penetration below plough depth.

NUTRIENT UPTAKE FROM FIELD SOIL

In most agricultural situations in the UK the topsoil layer is the main zone of nutrient supply and uptake. For most annual crops, 80 per cent of the total active root surface is contained in this layer. Whilst some nutrients, notably potassium and sodium, may occur at high available levels in the subsoil, the efficiency of root exploration is generally much reduced below 30 cm and so the subsoil contribution to total uptake is small. The zone of nutrient enrichment by fertilisers on permanent grassland is often restricted to the top 10 cm in the absence of plough incorporation. Subsoil nitrogen can be used where leaching has moved mineral nitrogen down the soil profile. For many spring-sown crops, the quantity of root is not sufficient for efficient use of subsoil nitrogen. If the soil moisture deficit is high enough for the main water uptake to be from the subsoil, the apparent effectiveness of subsoil nitrogen will be increased when compared with nitrogen in dry topsoil.

Work at the Letcombe Laboratory has shown that enrichment of part of the rooting zone with nitrogen or phosphorus can result in much more root growth in that area (see Plate 2.1). Thus the plant root system can respond to higher nutrient regimes. Recent work carried out to see whether nutrient enrichment of subsoils with phosphate and potash would encourage more roots and higher yields has not given economically worthwhile benefits.

MECHANISMS OF ROOT UPTAKE

Plants can take in nutrients along much of their root length but the most active area of uptake is near the root tips. Root hairs are particularly important to nutrient uptake because they greatly increase the area of active root surface and the volume of soil in close proximity to the root.

Around the plant root itself is the rhizosphere. This zone extends 1–2 mm from the root surface. It contains much organic matter due to sloughed root material and root exudates. Thus considerable microbiological activity occurs here. This activity is important for the cycling and solubilisation of phosphate.

Nutrients can reach the root surface by three main mechanisms:

1. root interception,
2. mass flow,
3. diffusion.

The quantity of available nutrient which the growing root comes into contact with is very small and thus root interception is the least important mechanism by which nutrients reach the root surface.

Mass flow is the process of nutrient movement to the root surface, occurring when soil solution moves to replace that taken up by the roots. The rate of nutrient supplied depends on the concentration in the soil solution and the amount of water taken up by the plant. It also depends on soil dryness and the speed with which water moves to the root surface.

Diffusion occurs without water movement when the nutrient concentration at the root surface is lower than that of the bulk soil solution. By thermal motion, nutrients will move towards the low concentration point until equilibrium is achieved. The ability of a soil to maintain an adequate nutrient concentration at the root surface will depend on the quantity of available nutrient in that soil and the rate of demand by the crop. Diffusion can only occur if soil moisture content is adequate to maintain water films between the root and the soil particles.

Diffusion is the dominant process of phosphate and potassium supply to the root. All nutrients move by diffusion under low transpiration rate conditions. When water uptake is high, calcium and nitrate are supplied predominantly by mass flow. If uptake is faster than supply, soil depletion of phosphorus and potassium can occur near the root surface. Phosphate depletion is a particular problem because of its lack of mobility.

As soon as a soil starts to dry out, mass flow and diffusion slow down and nutrient supply to the roots is reduced. This is because the size and number of liquid contacts between soil and root are reduced, and the air space increases. The practical result is that nutrient uptake is often restricted before water stress limits plant growth. This is particularly important for crop growth if young crops with a very high nutrient demand come under early water stress.

NUTRIENT MOVEMENT IN THE PLANT

Detail on the mechanisms of nutrient movement from the root surface to the xylem is outside the scope of this book. Suffice to say that anions in particular are taken up selectively and may be accumulated against a concentration gradient. This ion uptake is in part an active process requiring biochemical energy.

The main conducting tissues in the plant are the xylem and the phloem. Water and nutrients are moved from the roots to the upper parts of the plant along the xylem. The phloem transports mainly organic compounds in both directions.

Once nutrients have entered the xylem, they are quickly transported upwards in the transpiration stream. Nitrogen is transported as either NO_3^-, NH_4^+ or as amino acids. Thus the actively growing parts of the plant are kept well supplied with nutrients as long as the supply to the root surface is adequate.

The ability of the phloem to transport individual nutrients determines the extent to which a nutrient can be redistributed from one part of the plant to another. In particular it determines whether nutrients in older leaves can be reused in the younger parts of the plant. Potassium and magnesium are easily translocated in the phloem and are consequently found in high concentration in phloem supplied organs such as fruits and tubers. By contrast calcium translocation in the phloem is extremely small. Thus tissue with a low transpiration rate is inherently low in calcium, which may lead to calcium deficiency disorders in some fruits.

NUTRIENT DEFICIENCIES

Deficiency of a major nutrient causing visual symptoms is sometimes seen in field crops in the UK. When deficiency symptoms occur the cause is not always due to a low level of available nutrient in the soil. Sometimes other physical or biological factors reduce root uptake, producing deficiency symptoms. Manganese is the most common trace element deficiency seen in UK field crops.

Other factors which may cause poor root growth and/or nutrient uptake are soil compaction, drought and soil-borne pests and diseases. Various migratory and cyst nematodes can reduce root activity, as can root diseases such as take-all and club root.

Visual symptoms of common deficiencies are discussed in the appropriate chapters. The mobility of nutrients within the plant determines whether deficiency is first seen on young or old leaves. The mobile nutrients nitrogen, phosphorus, potassium, magnesium and manganese generally show symptoms on the older leaves first. Symptoms of calcium, boron and iron are first seen on the younger leaves or the growing point.

Visual symptoms of deficiency can often be confirmed by leaf analysis. This is very satisfactory for some elements but less sensitive for others. This is partly due to the change in critical concentration with leaf age. For some nutrients, notably iron, total leaf content is not correlated with metabolic activity.

Enzyme assay has been used to assist in the diagnosis of iron deficiency for which leaf analysis is unsatisfactory. The level of peroxidase is correlated with the degree of iron deficiency. However, the diagnosis is not iron specific and not applicable to all species. The need for fresh plant material adds to the reasons why the technique is not widely used in the UK.

Numerous studies have shown that one cation will antagonise the uptake of others, maintaining a similar total cation uptake. Various experiments have shown antagonism between calcium, ammonium–N (NH_4), potassium, magnesium and sodium. Any other cation will tend to reduce uptake of the dominant cation, calcium. This becomes a problem when calcium-related disorders are a risk (Chapter 3). High potassium is most likely to reduce the uptake of both calcium and magnesium. This may induce magnesium deficiency if the magnesium supply is low. High sodium following sea flooding can have a similar effect. Potassium uptake is not usually reduced by other cations.

Chapter 3

LIMING AND pH LEVELS

THE NUMEROUS marl and chalk pits in fields on soft, calcareous geological formations are a legacy of several centuries of liming. By the sixteenth century, harder limestones were being used in agriculture as burnt lime. Only during this century has machinery become available to grind these limestones for direct application. Many of our soils, notably the Chalks, Limestones and Chalky Boulder Clays, contain appreciable free calcium carbonate in the topsoil and will not need liming for decades or indeed centuries. Probably one-third of agricultural soils in England and Wales fall into this category. The remainder need regular liming.

The quantity of liming materials used annually in England and Wales has varied considerably over the last one hundred years. Accurate figures only exist for 1937–76 when a direct subsidy on liming materials was paid by the Government (Table 3.1). Various estimates of the annual lime lost from soils in England and Wales are between 3.3 and 3.7 million tonnes per annum.

While survey data and advisory experience indicate that arable farmers are maintaining their soil pH at an adequate level, grassland in England and Wales is receiving inadequate lime to maintain soil pH levels. Ground chalk or limestone currently costs around £12/tonne delivered and spread, but the considerable variation in price depends largely on distance from the quarry. This contrasts with a price of about £2.50/tonne in 1973.

CALCIUM

Calcium is a major nutrient required by all crops. If soil pH is maintained at a satisfactory level by liming, the supply of calcium as a nutrient is satisfactory for all crops. Calcium is required mainly in cell elongation and cell division. Calcium deficiency shows in the youngest

Table 3.1. Lime use in England and Wales

Year	Annual use (million tonnes $CaCO_3$)
1981	about 4.0
1976	4.3
1971	4.2
1966	5.4
1961	5.9
1956	7.1
1951	5.4
1946	3.3
1939	1.9
pre 1939	about 0.5

Source: Agricultural Lime Producers Council.

plant tissue resulting in stunting of stems and a lack of leaf expansion. In practice, the lack of mobility of calcium in the plant produces far more crop problems than absolute deficiency. Potatoes grown under very acid conditions are one of the few crops to produce field symptoms of calcium deficiency.

Calcium-Related Crop Disorders
Many crops produce disorders due to a lack of mobility of calcium within the plant. This is due mainly to the lack of phloem transport of calcium. Virtually all calcium movement is restricted to the xylem. Thus those parts of the plant with a low transpiration rate may not receive an adequate calcium supply, although the whole plant contains adequate calcium. Frequently the leaves are full of calcium but other

Table 3.2. Common calcium-related crop disorders

Crop	Disorder
Strawberry	leaf tipburn
Sprout	internal browning
Cabbage	internal tipburn
Lettuce	tipburn
Tomato	blossom end rot
Celery	black heart
Chicory	black heart
Apple	bitter pit
Tulip	topple

parts of the plant are deficient. Table 3.2 lists some of the common calcium-related disorders of UK crops. These generally occur when the water supply to the crop is intermittent. Soil pH is often satisfactory. Treatment with calcium foliar sprays is worthwhile for some disorders. Calcium chloride sprays are widely used on apples to reduce the incidence of bitter pit and other calcium-related disorders, many of which develop during storage.

SOIL ACIDITY AND pH

The pH of a soil depends on the amount of exchangeable calcium plus magnesium and free calcium carbonate present. In most soils in England and Wales, calcium is the dominant ion on the exchange surfaces, preventing hydrogen ions from causing low pH and acidity. The full pH scale is 1–14 with 7 being neutral. Lower figures are acid and higher figures alkaline. Well-aerated calcareous soils have a maximum pH of around 8.3 unless they contain free sodium carbonate when figures up to 10 can be achieved. This is restricted in temperate climates to sea flooded land. At the lower end of the range, soil pH does not drop much below 4.0 unless free acid, such as sulphuric acid, is present when figures down to 2.0 or lower may result.

The pH of a solution may be defined in precise terms as the negative logarithm of the hydrogen ion activity, but the pH of a soil suspension is a more complex property. Soil particles carry ions on their surfaces but these ions are not uniformly distributed throughout the suspension. Due to the concentration gradient between the particle surface and the bulk solution, the pH will be lower at the particle surface than in the surrounding solution. The higher the ionic concentration of the solution, the more similar the two values.

Generally pH is measured in distilled water at a ratio of 1 volume of dried and 2 mm ground soil to 2.5 volumes of water (because of the factors discussed above, a specified ratio is desirable). For some applications, measurement in a dilute salt solution such as 0.1 molar (M) calcium chloride is used to reduce the variation caused by fertiliser ions in the soil. This technique has not been generally adopted in England and Wales.

Other factors that may complicate the measurement of soil pH are changes in the oxidation/reduction state generally associated with waterlogging and changes in carbon dioxide concentration in the soil air. The first factor is of practical importance in acid sulphate soils considered later; the second factor is not important where standard laboratory pH measurement is employed.

PLANT GROWTH AND pH

The main influence of soil pH on plant growth is on the concentration of ions in the soil solution available for uptake by the plant roots. Table 3.3 gives the optimum pH levels to which mineral and peat soils should be limed for different rotations. The lower figures for organic soils are advised to reduce the risk of manganese deficiency on these very susceptible soils.

Table 3.3. Optimum pH levels (lowland soils)

	Arable crops	Permanent grassland
Mineral soils	6.5	6.0
Peaty soils	5.8	5.3

Source: ADAS.

Acidity restricts the growth of most crops due to the uptake of toxic amounts of available aluminium, manganese or iron. On acid soils the availability of calcium, bicarbonate and molybdate is reduced. Aluminium toxicity is the most common reason for soil acidity reducing plant growth. Barley and sugar beet are particularly susceptible to aluminium toxicity while potatoes are very tolerant. Under acid conditions crop uptake of manganese is invariably increased.

Calcium uptake is reduced under acid conditions but calcium deficiency as the primary cause of acidity on crop growth is only seen in potatoes and a few other species tolerant of high levels of both manganese and aluminium.

Permanent grassland shows considerable variation in its dominant species due to pH. Generally, low pH favours fescues as opposed to the more productive ryegrasses and clovers. However there are strains of wild white clover adapted to both low and high pH conditions. Hill pastures at low pH are commonly Nardus and Agrostis dominant, which are less palatable than the ryegrasses and Poas encouraged by raising the pH. At pH levels above 7.5 reduction in uptake of phosphate, manganese and boron may demand changes in fertiliser use on particular crops. The naturally high pH of many soils in England and Wales does not generally limit their usefulness for crop production.

At pH levels below 4.0 direct root injury may occur. Root development into subsoils when the pH is below this level is unlikely even if other toxicities are not an overriding factor.

CROP YIELD AND pH

The best source of data on the effect of soil pH on yield of arable crops is the Rothamsted work carried out over the period 1965–75 at both Rothamsted on clay-with-flints and at Woburn on sandy loam. In these experiments a range of pH levels was maintained for several years and different crops were grown. Examples are given in Table 3.4. Barley and spring beans were shown to grow best at 6.5–7.0, while yield of potatoes and oats was satisfactory above about 5.0.

Table 3.4. Range of pH and crop yield (t/ha)

Rothamsted clay-with-flints					
Barley 1965–7	pH	4.7	5.7	6.6	7.5
	yield	2.87	5.17	5.17	5.20
Potatoes 1974	pH	4.4	5.0	5.6	6.5
	yield	53	57	60	56
Oats 1975	pH	4.4	5.0	5.6	6.5
	yield	3.00	3.22	3.30	3.21
Woburn sandy loam					
Barley 1965–7	pH	5.2	6.4	7.2	7.4
	yield	4.93	5.06	5.20	5.17
Potatoes 1974	pH	4.8	5.4	6.3	6.7
	yield	59	58	53	56
Oats 1975	pH	4.8	5.4	6.3	6.7
	yield	2.27	2.71	2.44	2.41

Source: Rothamsted Experimental Station.

LIME LOSSES FROM SOILS

Under the UK conditions, any calcium or magnesium ions displaced by hydrogen are leached with an associated anion. Thus all soils are subject to calcium loss and will become more acid unless buffered by a reservoir of free calcium carbonate. Only a small amount of calcium is removed in crops.

The main soil processes that result in acidification are the oxidation of deposited atmospheric sulphur dioxide and the breakdown of soil organic matter. The main equations are:

$$\text{Atmospheric } SO_2 + H_2O + \tfrac{1}{2}O_2 \rightarrow 2H^+ + SO_4^{2-}$$
$$\text{Atmospheric } 2NO_2 + H_2O + \tfrac{1}{2}O_2 \rightarrow 2H^+ + 2NO_3^-$$

$$\text{Organic C} \rightarrow CO_2 + H_2O \rightarrow H^+ + HCO_3^-$$
$$\text{Organic N} \rightarrow NH_3 + 2O_2 \rightarrow H^+ + H_2O + NO_3^-$$
$$\text{Organic S} \rightarrow H_2S + 2O_2 \rightarrow 2H^+ + SO_4^{2-}$$

Considerable calcium is lost by leaching as bicarbonate, particularly in high pH soils. Work at Rothamsted has shown the lime losses over the pH range 5.0–8.0 double for each increase of one unit in pH. Table 3.5 shows figures for arable soil with 250 mm of winter leaching. Fertiliser use and the amount of winter rainfall will have appreciable effects on actual field losses. The higher the desired pH the greater the amounts of lime needed to maintain it.

Table 3.5. Effect of pH on annual lime loss

pH	Lime loss kg/ha $CaCO_3$
5.0	120
6.0	240
7.0	480
8.0	960

Source: Rothamsted Experimental Station.

Effect of Fertilisers

Any nitrate, sulphate or chloride ions added to the soil in fertiliser will be leached from the soil if not taken up by the crop or utilised by the soil micro-organisms. Most sulphate and chloride will be lost, with an associated calcium or magnesium cation. The amount of nitrate lost will vary: very little will be leached from grassland, more is likely to be lost from arable land particularly if crop uptake is restricted for any reason.

Ammonium-containing fertilisers have a greater acidifying effect as shown by the equation below:

$$2NH_4^+ + 4O_2 \rightarrow 2NO_3^- + 4H^+ + 2H_2O$$

All ammonium ions added to soil will be oxidised to nitrate unless taken up by the crop or utilised by micro-organisms, resulting in considerable release of hydrogen ions which in turn will displace calcium. Ammonium sulphate is notorious as an acidifying fertiliser because its associated anion is lost by leaching in addition to the

**Table 3.6. Amount of CaCO₃ needed to neutralise the
effect of various fertilisers**

	$CaCO_3$ kg
1 kg N as	
Ammonium sulphate (21% N)	5 –7
Ammonium nitrate (34% N)	} 2 –3
Urea (45% N) ammonia	
Calcium ammonium nitrate (26% N)	$1\frac{1}{2}$–$2\frac{1}{2}$
Calcium nitrate (15% N)	Nil
1 kg K₂O as	
Potassium sulphate (50% K₂O)	} 1
Potassium chloride (60% K₂O)	

ammonium effect. Table 3.6 gives figures for the likely acidifying effects of various fertilisers; calcium nitrate may give a slight increase in pH, phosphates have very little effect on soil pH. High rates of animal manures may cause an increase.

Crop Removal
The amount of calcium removed in crops is small as shown by the figures in Table 3.7 expressed in calcium carbonate equivalents. High calcium contents are restricted to forage and vegetable crops where high dry matter yields of leaf are removed from the field. When crop residues are burnt, most of the calcium will remain in the ash.

Individual Field Lime Losses
The combination of these factors to calculate total lime loss from an individual field over a particular period of time has not been reliably

Table 3.7. Crop removal of calcium

Crop	Yield t/ha	Removal (kg/ha $CaCO_3$ equivalent)
Cereal grain		
plus straw	7	50
Sugar beet – roots	40	60
Potatoes	40	20
Oilseed rape	3	30
Grass – silage	50	150
– hay	5	40
Kale	60	450

achieved to date and needs more development work. Recent data from Rothamsted indicate that total annual lime losses from arable land vary from 0.5 to 1.0 tonnes per hectare (t/ha) $CaCO_3$ per year. The losses will be greatest on high pH sandy soils with high nitrogen fertiliser usage and high excess winter rainfall. Loss from grassland will generally be less in comparable winter rainfall situations. Losses following ploughing out of grass can be very high while the organic matter is breaking down and nitrogen is being released.

CROP SUSCEPTIBILITY TO ACIDITY

Individual crops vary considerably in the soil pH level at which acidity begins to reduce yield. If visual symptoms of poor growth are seen in a crop due to acidity, yield loss is probable. In practice it is necessary to

Table 3.8. Minimum satisfactory field pH levels for mineral soils

pH level	Crop
7.0	brassicas (clubroot risk high)
6.5	barley, beans, sugar beet, mangolds, peas, red clover, lucerne, asparagus, redbeet, celery, leeks, lettuce, mint, spinach, tulip, blackcurrant
6.0	kale, linseed, maize, mustard, oilseed rape, oats, swede, turnip, wheat, white clover, cocksfoot, timothy, brussels sprout, cabbage, carrot, cauliflower, onion, parsnip, rhubarb, sweet corn, daffodil, pear, plum, raspberry, redcurrant, hops
5.5	potatoes, rye, chicory, parsley, apple, blackberry, strawberry
5.0	wild white clover, fescues, ryegrass

maintain the field pH at a level high enough to satisfy the most sensitive crop in the rotation.

Table 3.8 gives a guide to satisfactory field pH levels for the major crop plants grown in England and Wales. Levels of pH 0.5 units below those suggested are likely to restrict growth on mineral soils, particularly under dry soil conditions. Where figures below these are found in patches in a growing crop, topdressing with lime should be considered.

Acidity Diagnosis

Generally crops suffering from acidity appear stunted and chlorotic (yellow). Field occurrence is usually in patches often associated with areas of coarser textured soil (see Plate C.1). Those crops sensitive to aluminum toxicity, particularly cereals and sugar beet, show chlorotic leaf tips or margins and stubby, highly branched root systems. The latter is particularly characteristic (see Plate 3.1). Often phosphate uptake is limited, giving some purpling in cereals. In contrast potatoes grown at low pH show calcium deficiency symptoms of restricted shoot growth and very small younger leaves.

Apples develop manganese toxicity under acid conditions which shows as measly bark, while some varieties of strawberries show purple veining. The latter symptom of purple or browning of veins or spots adjacent to the veins is characteristic of manganese toxicity.

While crop symptoms are a useful guide to the likelihood and distribution of a crop acidity problem in the field, soil pH measurement is invariably the most useful diagnostic technique to confirm the cause of the poor growth. Plant analysis is generally restricted to special cases, more commonly used for fruit than in annual crops. On some soil types it may be necessary to check the pH of the subsoil as well as the topsoil. The most common problems of this type are on lowland peat soils and sandy heathland soils recently reclaimed for agriculture.

MEASUREMENT OF SOIL pH

While accurate measurement is always desirable, pH generally varies considerably both with depth and in different areas of the field. Thus careful depth sampling followed by bulking of the individual soil cores for precise laboratory analysis will be accurate, but it may be misleading.

When sampling most potentially acid soil types it is sensible to use a coloured indicator to test each core. In practice liming must be carried out to correct the lowest pH in the field rather than the average from a bulked sample. It is not uncommon for laboratory soil analysis to be blamed for giving the wrong result when acid patches occur in the next crop. This can easily happen if soil in part of the field contains free calcium carbonate, and thus gives a satisfactory pH on a bulked sample. Problems may also arise in the measurement of pH on recently limed soils. If the laboratory figure is read after shaking the soil in water, the lime will have fully reacted. This reaction may take several months to occur in the field.

Various portable pH meters are commercially available. Those using a glass electrode system can be as good as a laboratory analysis as long as sample preparation is thorough and the instrument is well maintained and calibrated. The spur type instruments which are pushed into the soil are generally less reliable as they are very dependent on soil moisture content. In the main, indicator testing is more reliable for field assessments.

Plate 3.1. Cereal root system grown in acid soil   *Agricultural Lime Producers Council*

Using a Soil Indicator
The ADAS soil indicator is a mixture of bromothymol blue and methyl red. It can be used to assess most soils to the nearest 0.5 pH unit. While use of sample tubes with barium sulphate to clear the suspension can be used, dish testing is adequate for most purposes (see Plate 3.2).

Technique:

1. Clean a white dish with distilled water or a little indicator.
2. Place a small quantity of the soil to be tested in the dish and add enough indicator to saturate it.
3. Gently swill the mixture, avoiding breakdown of the aggregates. Very gentle action is needed for fine sandy and light silty soils to avoid making a cloudy suspension.
4. After 1–2 mins assess the colour of the indicator in the dish, using the colour interpretation card (see Plate C.2).

Problems of colour adsorption occur on some organic soils. Highly coloured soils may mask the colour change by giving an orange or reddish suspension of soil particles. If this happens as soon as indicator is added, the test should not be relied upon. Problems also occur due to colour blindness of the operator or use of old indicator. After twelve months, the supply should be laboratory checked or discarded. Contamination may also cause problems. Contact with the fingers while shaking will decrease the pH. A drop of indicator on the skin goes orange or red. Also avoid stirring with twigs or straw and scrape burnt straw ash from the soil surface before sampling.

OTHER EFFECTS OF pH

As well as influencing the chemistry of the soil, pH and the presence of free calcium carbonate influences various physical and biological aspects of soils. The structure and workability of calcareous clays are generally better than non-calcareous clays. On some very unstable silty clays, high applications of liming materials may be justified to improve the stability even though the pH is already 6.5–7.0. Due to the cost and risk of inducing trace element deficiencies, this treatment is limited in practice to difficult patches in fields, usually on heavy marine silts.

Acidity reduces the activity of much of the soil flora and fauna. The rate of organic matter decomposition is much slower under acid conditions. Acid permanent grassland frequently builds up a thick surface root mat because of a lack of biological activity. Worm activity is much less under acid conditions.

Some important soil-borne plant diseases are influenced by pH and liming. Club root of brassicas is less active at high pH. Recent work has confirmed that liming to pH 7.5 or above is a fairly effective control for this disease. In contrast, common scab of potatoes is invariably made worse by recent liming. An increase in surface acidity will increase the adsorption of most soil acting herbicides. This results in much poorer weed control at the normal application rates.

LIMING

The principle of liming a soil is to apply a material which will react with the soil, removing hydrogen ions and replacing them with calcium or magnesium. Calcium carbonate is the normal material used and reacts as shown by the equation:

$$CaCO_3 + 2H^+ \rightarrow Ca^{2+} + CO_2 + H_2O$$

Calcium hydroxide and the silicates in basic slag behave similarly:

$$Ca(OH)_2 + 2H^+ \rightarrow Ca^{2+} + 2H_2O$$
$$CaSiO_3 + 2H^+ \rightarrow Ca^{2+} + SiO_2 + H_2O$$

Plate 3.2. The use of soil pH indicator

A range of pH can be found on all soil textures. The amount of lime needed to bring the pH up to the figure required for the rotation depends mainly on the soil cation exchange capacity (CEC). The higher the CEC the more liming material is needed to neutralise the soil acidity and produce the desired rise in pH. The main soil components determining CEC are clay and organic matter content. Thus organic and clay textures need far more lime than sandy textures to give a similar increase in pH.

ASSESSMENT OF LIME REQUIREMENT

ADAS has recently changed from a laboratory method of measuring soil lime requirement using para-nitrophenol buffer to a method based on soil texture and pH. This change was made because the two methods show a good linear relationship and also because the laboratory method can present practical problems in measuring small lime requirements and in converting the laboratory figure to a field recommendation.

The current method aims to achieve a pH of 6.5 for arable soils to a depth of 20 cm. The recommendation is given in tonnes per hectare (t/ha) of ground chalk or limestone. This figure assumes that the lime is cultivated into the soil and that there is adequate moisture for the reaction to take place. Unless rotavated into the soil, mixing in practice will be less complete. Variation within the topsoil often lasts for 1–2 years after heavy liming even though ploughed annually. The recommendation does not take into account lime losses over the next few years.

Lime requirements for lowland grass are based on soil texture and pH for a soil depth of 15 cm. The maximum application rate recommended is 7 t/ha. In the absence of cultivations, this is the maximum amount needed to affect soil pH over a period of 2 to 3 years. Higher rates would only influence the top few centimetres and may adversely affect the nutrient balance of the herbage.

Table 3.9. Lime requirements for arable soils (20 cm depth) to achieve pH 6.5 (5.8 for peats) and grassland (15 cm depth) to achieve pH 6.0 (5.3 for peats)

pH level	Sands		Loams		Clays		Peats	
				t/ha $CaCO_3$				
	Arable	Grass	Arable	Grass	Arable	Grass	Arable	Grass
6.5	0	0	0	0	0	0	0	0
6.0	4	0	5	0	6	0	0	0
5.5	7	3	8	4	10	4	8	0
5.0	10	5	12	6	14	7	16	6
4.5	13	7	15	7	18	7	24	7
4.0	16	7	19	7	22	7	32	7

Table 3.9 shows the amount of lime needed to correct acidity depending on soil pH and texture. The arable figures may need correction if plough depth is more or less than 20 cm. It indicates the variation in lime requirement between soil types and can be used when only pH values are available. The table should not be extrapolated below pH 4.0.

Laboratory assessment of the lime requirement of peat soils is not very satisfactory. Normal practice is to lime at rates indicated by field experiments. The lower optimum pH levels for peat soils balance the lower risks of manganese and aluminium toxicity against the greater risk of manganese deficiency, when compared with mineral soils at the same pH.

Levels of pH below 4.0 are generally restricted to the subsoils of acid sulphate soils. These occur mainly in ground water situations and present low pH problems after drainage. This allows aeration and thus oxidation of sulphides (mainly pyrites) resulting in the production of sulphuric acid. It is impossible to assess the lime requirements of these situations accurately until oxidation is complete. Then a laboratory titration for exchangeable acidity can be used. These soils often need huge applications of lime—sometimes in excess of 250 t/ha—to neutralise the acid produced. They may take decades to become productive soils. Other shallow, peaty soils may have a 'drummy' acid layer below pH 4.0 at the junction with the mineral subsoil. These can generally be improved at an acceptable cost.

The field pH requirement for a particular crop may be different to the general arable and grass figures suggested. If this is so, the lime requirement figure should be recalculated to meet the particular figure required. Upland grass liming experiments generally show that liming above pH 5.0 will not be worth while, unless reseeding is also carried out as part of the improvement. Land in continuous early potatoes need not be limed above pH 5.0–5.5. At the other extreme, land in continuous brassicas may justify the pH being maintained at 7.0–7.5 to reduce the clubroot disease risk.

LIMING MATERIALS

Plentiful supplies of calcium and magnesium carbonates as chalk or limestone occur over most of England and Wales. Mid Wales and south west England are the furthest from major deposits (Figure 3.1).

A wide range of alkaline materials can be used to neutralise the effects of soil acidity; in practice the choice is economically restricted to a few. The definitions and quality characteristics with which liming materials sold for agricultural use in Great Britain must comply, are detailed in the 1977 Fertiliser Regulations (Statutory Instrument No 1489).

For all materials the main requirement is a declaration of its neutralising value (NV). This figure is a comparison in percentage with pure calcium oxide at 100. Most liming recommendations are given in rates of calcium carbonate (NV 50).

When comparing the prices of different liming materials spread on the field, it is important to ensure that the products are compared on an effective NV basis. If a higher rate is needed because of particle size or a material takes longer to spread, what first appears cheap may not be the best, most agronomically effective buy.

One tonne of ground limestone (NV 50) is equivalent to $\frac{50}{70}$ t of hydrated lime (NV 70). The following materials are the main products sold in England and Wales together with approximate NVs and their main agronomic characteristics (Table 3.10). Over half the material sold is ground limestone; a further quarter is ground chalk or waste

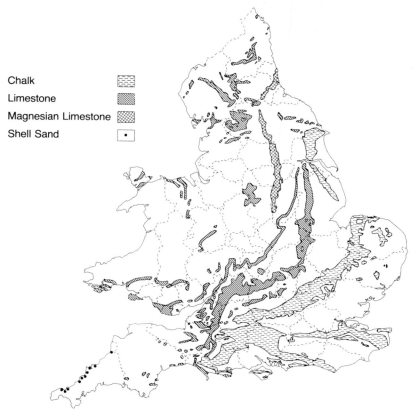

Figure 3.1. Main deposits of liming materials in England and Wales
Source: ADAS

Table 3.10. Main UK liming materials

	Approx neutralising value (% NV)
Ground limestone	50
Screened limestone and limestone dust	48
Ground magnesian limestone	50–55
Ground chalk	50
Screened chalk	45
Calcareous sand: shell sand	30–40
Sugar beet factory lime sludge	20–25
Water works lime sludge	20–25
Burnt lime	80–95
Hydrated lime	70

Plate 3.3. Chalk spreading   *Pertwee Landforce Ltd*

lime. Rather less than 20 per cent is composed of numerous other materials. Historically basic slag has been an important source of lime but this is no longer available in appreciable quantities.

Ground limestone (NV about 50)
This is finely ground naturally occurring limestone, composed of almost pure calcium carbonate. All the sample must pass through a 5 mm sieve and at least 40 per cent must pass through a 150 micron sieve. Experiments have shown no agronomic advantage from grinding finer than 55 per cent through 150 microns.

Screened limestone and limestone dust (NV about 48)
By-products of roadstone production, these materials require a minimum of 20 per cent to pass through a 150 micron sieve. The coarser particles, particularly in screened limestone, may take several years to react. Consequently these products may need higher rates than indicated by NV to be agronomically as effective as ground limestone. Coarse versions of the above with a minimum of 15 per cent passing 150 microns are correspondingly less effective.

Ground magnesian limestones (NV about 50–55)
A similar range of magnesian limestones is available to the same grinding specifications as the calcitic limestones. They must contain not less than 15 per cent MgO (9 per cent Mg). The magnesium is slowly available and the large quantity applied especially over a number of years will maintain a satisfactory soil magnesium level on all soils.

Ground chalk (NV about 50)
This is natural chalk ground so that 98 per cent will pass a 6.3 mm sieve. This statutory requirement is satisfactory for soft chalks, but the harder chalks of northern England need finer grinding to be equally as effective. Chalk spreading is shown in Plate 3.3.

Screened chalk (NV around 45)
Commonly produced in south and east England, the statutory requirement is for 98 per cent to pass a 45 mm screen. In practice it is usually ground to pass a 6.3 mm, 12.7 mm or 25.4 mm sieve. These products are all equally efficient in the long term once the lumps have weathered. Speed of action is increased considerably if the material is on or near the soil surface during the first winter after application.

Calcareous sand—shell sand (NV 30–40)
Collected mainly from the north Cornish beaches, the shell content of these materials makes them effective for local use; 100 per cent must pass a 6.3 mm sieve.

Sugar beet factory waste lime (NV depends on water content)
Large quantities of waste lime are produced at sugar beet factories. It generally contains 50 per cent and sometimes over 60 per cent water. Thus rates of two to three times those of ground chalk or limestone are needed. It is a precipitated product and therefore is in a very finely divided form compared with ground products. It has been shown to be particularly effective on peat soils in the Fens. Spreading is its major problem; it is best handled with muck spreading machinery (see Plate 3.4). It contains some nitrogen, phosphate and magnesium: average figures are 25 kg N, 80 kg P_2O_5 and 35 kg Mg per 10 tonnes.

Water works waste
Available in some areas, it is similar to beet factory sludge being variable in water content unless dried. It may contain some hydrated lime.

Plate 3.4. Sugar beet factory waste lime   *British Sugar plc*

Burnt lime
Burnt lime, ground burnt lime, hydrated lime and kibbled burnt lime
are all effective liming materials with NVs of 70 or above. They are
generally too expensive to be worth considering for agricultural use.

PRACTICAL ASPECTS OF LIMING

When farming potentially acid soils it is necessary to develop a liming
policy. This should ensure no yield loss due to acidity and that the cost
of liming does not all arise in one year.

Annual Crops
For annual cropping, liming should be carried out every four to seven
years to maintain a satisfactory pH level. The frequency may need to
be greater in some high lime loss situations. Where a sensitive crop
such as sugar beet occurs in the rotation, it makes sense to lime before
the preceding crop so the lime is ploughed back up for the beet. It will
still be necessary to check the topsoil pH on very sandy soils after a wet
winter.

The rate required assuming 0.5–1.0 t/ha loss per year will be 2.5–
5 t/ha every five years. Soil analysis will give a more precise requirement where a low pH is encountered.

It is satisfactory to plough down a rotational application of lime as long as the surface soil pH is not too low for the next crop. If the pH is low, the lime must be applied after ploughing so that the surface soil has a suitable pH for seedling growth. A constant depth of ploughing should be maintained. Where pH is likely to limit the yield of the next crop, autumn liming will be more effective especially if cultivated in prior to drilling. If surface acidity has built up under direct drilling or minimum cultivations on a naturally calcareous soil it may be simpler to plough and bring up calcareous soil rather than apply lime.

If acidity is diagnosed in a young growing crop, topdressing with 3–5 t/ha of finely ground chalk or limestone will be partially effective if followed by rain. Hydrated lime is often recommended for topdressing but it is about four times as expensive as ground chalk or limestone and only more effective if there is sufficient rain or irrigation is applied.

Where severe acidity occurs as commonly found with ploughed up permanent grass, it may be necessary to avoid acid sensitive crops for two years. In any severe acidity situation, half the topsoil lime requirement should be ploughed down followed by the other half after ploughing. If the subsoil is acid, a greater amount of lime should be applied which will at least help to neutralise it. Deep incorporation of lime is generally needed on acid peats if the subsoil pH is to be improved.

The time taken for a lime application to have maximum effect on soil pH depends mainly on the effectiveness of mixing, fineness of grinding of hard limestones and soil moisture. Three months should be adequate for low application rates, but allow one to two years where severe acidity is being corrected.

Perennial Crops

As acidity develops from the soil surface in cultivated land that has previously had a satisfactory pH, the top few centimetres (cm) should be sampled and tested to get an early warning of acidity. Once diagnosed, liming 'little and often' is the best policy. If severe acidity has developed, it is very slow to reverse without cultivation. This is a particular problem in crops under herbicide management such as fruit and hops.

Acid grassland will improve more quickly if lime can be harrowed into the surface mat. If the mat is left undisturbed, it may absorb the lime added with no benefit to soil pH. If the top 7.5 cm is acid, the standard 15 cm depth lime requirement figure on the soil analysis report should be applied. Before establishing any perennial crop, it is important to ensure that the soil has been adequately and effectively limed.

Chapter 4

NITROGEN

NITROGEN is the nutrient required in the greatest quantity by most crops. It is also one of the most complex in behaviour, occurring in soil, air and water in inorganic and organic forms. For this reason it provides the most difficult problem in making fertiliser recommendations for crops.

THE NITROGEN CYCLE

Figure 4.1 shows the main chemical forms of nitrogen in the soil–crop system and the directions in which changes may occur. Nearly all these transformations of nitrogen are carried out by soil micro-organisms. Whenever change in the soil organic constituents is taking place, nitrogen is being changed from one form to another.

The main forms in which nitrogen is added to the soil as inorganic fertilisers are:

Nitrate (NO_3)
Ammonium (NH_4^+)
Simple amides ($- NH_2$)

In addition, nitrogen is supplied in animal manures containing both ammonium and organic forms. Gaseous nitrogen (N_2) is fixed from the atmosphere.

One of the main problems in nitrogen research is accounting for all the nitrogen in a soil–plant system. Nitrogen can be lost by leaching through the soil to the drains or aquifer and also in gaseous form to the atmosphere. These losses are difficult to measure in field experiments. The following sections discuss each of these major transformations and their practical implications.

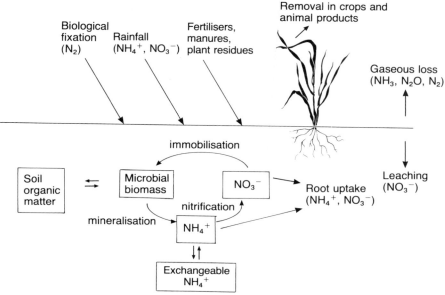

Figure 4.1. Nitrogen cycle

PLANT METABOLISM

Most crop plants can take up both ammonium and nitrate ions through the root system. Most uptake at normal soil pH levels for crop production is as nitrate. This is due to the rapid conversion of ammonium to nitrate in the soil following application of any ammonium fertilisers.

Within the plant, nitrate is first reduced by the enzymes nitrate reductase and nitrite reductase to ammonium. Further enzymes then convert ammonium to simple soluble organic molecules, amino acids, amides and amines. This stage is not reversible in crop plants. This is followed by conversion to proteins, the dominant nitrogen fraction in green plant material, and nucleic acids. Varying proportions of enzyme, storage and structural proteins occur, depending on the particular plant organ.

These nitrogen transformations occur at several places within the plant as the nitrogen is translocated from the roots to the youngest parts of the plant. Nitrogen is very mobile within the plant. Any shortage of nitrogen in young tissue is generally met by mobilising compounds from the older leaves. This results in a loss of chlorophyll showing as yellowing or chlorosis.

DIAGNOSIS OF DEFICIENCY

Nitrogen deficiency is characterised in almost all species by overall chlorosis of the older leaves (see Plate C.3). In severe cases the whole plant is stunted and leaves remain small. In many species, eg. brassicas, magnesium deficiency symptoms are also seen on the older leaves of nitrogen deficient plants. This occurs where chlorophyll has been broken down and nutrients have been mobilised to the younger parts of the plant.

Where recommended rates of nitrogen fertiliser have been applied and excessive leaching has not occurred, visual symptoms are generally indicative of poor root activity. The crop will not necessarily benefit from extra fertiliser nitrogen. During periods of high nitrogen demand when growth rates are high in April/May cereal crops commonly show yellowing of the leaf tips. This is generally related to moisture stress, take-all disease or some other limitation to effective nutrient uptake. Similar symptoms are associated with other soil-borne problems such as nematode attack on the roots. Lack of nitrogen uptake results in earlier maturity of many crops.

In other situations poor soil structural conditions combine with too much or too little rainfall to impair nutrient uptake by a restricted root system. Again nitrogen deficiency symptoms are commonly the first sign of trouble.

Where visual symptoms are not obvious, leaf analysis can be useful to decide whether the crop has taken up adequate nitrogen. The main problem is the inherent change in leaf nitrogen content with age. Unless sampling is specific to a particular leaf at a particular growth stage, leaf analysis should not be relied upon for making fertiliser decisions.

Measurement of the nitrate content of plants, either whole leaves or sap squeezed from the petiole, has been advocated by some researchers. If a young plant contains appreciable nitrate waiting to be metabolised, it is adequately supplied with nitrogen. If the level is low, this indicates a deficiency. Often leaf colour or leaf total nitrogen content will provide similar information. The sap nitrate test is the only one able to be quantitatively performed in the field using test strips. However assuming that a policy of nitrogen application based on experiments taken to yield is being followed, it is difficult to see any real value to the grower from this test. A further difficulty is that it is known that transient nitrogen deficiency early in the life of many crops does not necessarily produce a yield penalty at harvest. If a deficiency is diagnosed, it may be due to other factors than soil nitrogen supply so that extra nitrogen fertiliser may or may not be beneficial.

MINERALISATION

A wide range of heterotropic micro-organisms, using carbon as their energy source, can transform soil organic matter first to amino–N and then to ammonium–N. As this process is carried out by many soil bacteria and fungi, the main limitation to the rate of change is environmental rather than a lack of the right micro-organisms.

When organic materials are used by micro-organisms ammonium–N may be released or all the nitrogen mineralised may be incorporated into the organisms themselves as they multiply in the soil. If the particular organic matter source is very low in nitrogen content, the organisms will need extra nitrogen to grow. In this case inorganic nitrogen already present in the soil will be used up by the soil micro-organisms. This process is called nitrogen immobilisation.

Rates of microbial activity in the soil depend on several factors. Soil water content is important because as the soil dries, biological activity declines rapidly. Oxygen supply must also be adequate as soil organisms require oxygen to function. However if soil oxygen supply is low due to waterlogging, breakdown may still occur by bacteria which can function in the absence of oxygen.

Rates of breakdown of organic matter are slower under acid conditions than when the pH is 5 or above. This is mainly due to the restricted range of soil flora and fauna that exist under acid conditions. Temperature is the other major factor determining rates of breakdown. Over the range 10–30°C, an increase of 10°C increases the rate of microbial activity by two or three times.

SOIL ORGANIC MATTER

Mineral soils contain varying amounts of humus or soil organic matter. This is usually concentrated in the top 10 cm of undisturbed soils or distributed to plough depth in cultivated land. On cultivated land the level will depend on soil texture, the frequency of grass in the rotation and the application of organic manures. Typical levels are shown in Table 4.1 for mineral soils. By contrast, peat soils may contain up to 50 per cent by weight of organic matter. The rate of biological change of the soil organic matter and the quantity of mineral nitrogen released for crop uptake varies for different fractions of the total organic matter. The total nitrogen content of the soil is measured in t/ha—up to 10 t/ha is not uncommon. A small biological change can release agronomically significant amounts of mineral nitrogen.

When annual crop residues are returned to the soil, breakdown is normally 70 per cent complete within twelve months. There is a much greater variation in the initial rate of breakdown of different plant residues following incorporation. When organic matter is added to the soil the micro-organisms breaking it down can obtain their nutrient requirement from the material itself or from the soil. Nitrogen is the nutrient most likely to limit the rate of breakdown. The need for an external source of nitrate depends on the carbon:nitrogen ratio (C:N ratio) of the organic material. Materials rich in nitrogen, such as lucerne, will release mineral nitrogen as soon as breakdown starts. By contrast wheat straw is low in nitrogen and will immobilise soil mineral nitrogen during its breakdown. Work at Rothamsted quotes nitrogen contents of less than 1.2–1.3 per cent nitrogen (C:N of 30) causing immobilisation of soil or fertiliser mineral nitrogen and nitrogen contents above 1.8–2.0 per cent nitrogen (C:N of 20)

Table 4.1. Typical levels of organic matter in lowland mineral topsoils

Soil texture group	Arable	Ley/arable	Permanent grass
Sand	1.0–1.6	1.4–1.8	3–5
Light loam	1.4–2.5	1.8–3.0	4–6
Light silt	2.0–2.6	2.5–3.0	5–7
Medium loam ⎫ Medium silt ⎭	2.4–3.0	2.8–3.2	5–8
Clay	3.2–4.0	3.7–4.2	5–10

Source: ADAS.

resulting in fairly rapid mineralisation. High lignin content materials will break down more slowly than those consisting mainly of cellulose.

Most experiments indicate that 1 tonne of wheat straw requires about 10 kg nitrogen fertiliser addition to ensure rapid breakdown without immobilisation. The microbial demand for nitrogen will only compete directly with nitrogen supply to a growing crop if the mineral nitrogen reserves of the soil are insufficient. In most arable situations where crop residues are incorporated, the soil mineral nitrogen reserves are adequate to supply both plant and micro-organisms. Extra nitrogen is only likely to be needed for high rates of straw incorporation on sandy soils. Material incorporated in the spring is much more likely to justify extra nitrogen fertiliser than residues incorporated in the autumn when crop nitrogen demand is low.

Recent work has shown that the soil organic matter can be considered as two separate components. The biomass is the living microbial population which is very active and constantly turning over nitrogen as individual micro-organisms die and new ones appear. The rest of the soil organic matter is much more resistant to breakdown and is only metabolised very slowly. Peaty soils can be shown to release considerable quantities of mineral nitrogen. Mineral soils with less than 8 per cent organic matter release much smaller amounts of mineral nitrogen per year. It is hoped that current experimental work using nitrogen isotope labelling will help to quantify the amounts of mineral nitrogen available from the various soil organic matter components.

GREEN MANURING

Green manuring, the growing and ploughing in of a green crop, is a traditional technique used to conserve nitrogen and to help maintain soil organic matter levels. Its use for game cover is also a factor on some farms. The chosen crop for autumn use is usually mustard or forage rape. It is sown immediately after harvesting the main crop and ploughed in before Christmas.

The improvement in organic matter levels by this technique is very small. The main agronomic justification must come from conserving soil nitrogen. While small increases in soil nitrogen retention can be achieved on light soils, the benefits are unreliable and generally not worthwhile in current economic circumstances. In most cases an autumn-sown cereal or oilseed rape crop has largely replaced green manuring and can be expected to conserve soil nitrogen in crop residues at least as well as a ploughed-in green crop. Pest and disease problems preclude the use of brassica green manure crops in most sugar beet or oilseed rape rotations.

NITRIFICATION

This involves the conversion of ammonium–N first to nitrite–N and then to nitrate–N mediated by specific soil bacteria, *Nitrosomonas* and *Nitrobacter*.

$$NH_4^+ \xrightarrow{\text{Nitrosomonas}} NO_2^- \xrightarrow{\text{Nitrobacter}} NO_3^-$$

Since the original work identifying *Nitrosomonas* and *Nitrobacter*, a number of other bacteria have been shown to perform the first part of the transformation. As all field soils contain these bacteria, the change from ammonium to nitrate will take place as long as pH, temperature and moisture levels are satisfactory. Generally the second stage of the process is more rapid than the first so nitrite does not accumulate.

Below pH 5.0, rates of nitrification are much reduced. Nitrification does not occur in the absence of oxygen. As with most soil biological processes, rates of activity decline rapidly if the soil is waterlogged or too dry. Temperature has a marked effect on rates of nitrification. Over the soil temperature range 5–30°C, the process is rapid; however, nitrification occurs much more slowly below 5°C.

Various toxic materials in the soil can inhibit nitrification. Ammonia is toxic to both groups of organisms, thus high concentrations of ammonia fertiliser will reduce the rate of its subsequent conversion to nitrate. Other materials commonly added to soils which influence nitrification are the partial soil sterilants such as methyl bromide or dazomet. These commonly cause an initial decrease in nitrification, often followed by an increase later in the season as biological activity resumes again.

Nitrification may also be reduced by specific chemical inhibitors such as nitrapyrin (N-serve) and dicyandiamide (Didin, DCD). When added to the soil, both materials inhibit nitrification. When added to fertiliser, these materials reduce the rate of ammonium conversion to nitrate at soil temperatures above about 6°C. The only likelihood of an economic advantage comes when there is a need to apply nitrogen fertiliser earlier than normally recommended. No advantage is likely if fertiliser is applied at normal times. The length of time that the inhibitor delays nitrification depends mainly on temperature. Breakdown of these chemicals is much quicker in summer than in autumn or spring. There is more scope for economic benefits when dicyandiamide is added to slurry, particularly if the slurry is injected or rapidly incorporated into the soil. As with most slurry experimental work, results to date are variable but yield responses have been shown in some experiments where modest slurry rates have been applied in late autumn or early spring. Benefits are only likely where care is already being taken to try to maximise the nitrogen fertiliser value of slurry on the farm.

LEACHING

Nitrate is very mobile in the soil. For this reason it is susceptible to leaching down the soil profile when rain or irrigation falls on a soil already at field capacity. During the period of excess winter rainfall from October to April, depending on the area, any nitrate in the profile may be leached beyond rooting depth and lost by drainage to rivers or percolate towards the aquifer.

The extent of nitrate movement varies considerably depending on soil type. On sandy non-aggregated soils the process can be considered as one of displacement. As water falls on the soil surface, the water in the profile moves down like a piston, carrying dissolved ions with it. The distance moved down the profile will depend on how much water the soil holds at field capacity. On aggregated soils the process is more complex and generally the amount of nitrate leached is much reduced.

Table 4.2. Estimated percentage of nitrogen fertiliser application leached below 50 cm soil depth on sandy soils

Texture	Excess rainfall mm					
	25	*50*	*75*	*100*	*125*	*150*
Sand	< 10	23	38	49	58	62
Loamy sand	< 10	14	28	39	48	55
Sandy loam	< 10	< 10	17	28	37	44
Sandy silt loam	< 10	< 10	13	22	31	39

Source: based on NVRS leaching model.

On such soils, water does not move uniformly but follows the cracks and fissures between the aggregates. Much of the nitrate present is held within the aggregates and only slowly equilibrates with the soil solution. The result is a much less efficient removal of soluble nutrients from the soil.

Most practical problems of leaching loss of fertiliser occur in the spring on sandy soils. Recent work at Rothamsted and National Vegetable Research Station (NVRS) Wellesbourne has attempted to model the leaching losses of nitrate from soils. Table 4.2 is based on the NVRS model for sandy soils. The variation with soil texture is due to the different water holding capacities of the soils at field capacity. These figures assume complete nitrification of fertiliser applied and no crop uptake. They are mainly of use for spring-sown crops which often have a very shallow root system when heavy rain occurs in March or April.

GASEOUS LOSS

Direct loss of gaseous ammonia may occur following the application of inorganic fertilisers, particularly urea, and from organic manures. The mechanisms and agronomic consequences are discussed in Chapter 9 on types of fertilisers.

The other major loss that is notoriously difficult to measure in experiments is the microbiological denitrification of nitrate to gaseous nitrogen (N_2) and nitrous oxide (N_2O) which are lost to the atmosphere. This change is carried out by bacteria when the soil oxygen concentration is low. As it is an anaerobic process, it is normally associated with waterlogged soils. However it has been shown that well-drained clay soils can have oxygen-free conditions within aggregates, without the whole soil being waterlogged. This occurs if the soil oxygen demand by the micro-organisms is high, as occurs when the soil is warm and contains a lot of recently incorporated crop residue.

Losses are generally low during the winter but the potential loss increases greatly when soil temperatures rise above 8°C. Nearly all the loss occurs from the topsoil which has the most biological activity. Losses depend much on the physical state of the topsoil and the stage of breakdown of residues from the previous crop. Fields following late harvested sugar beet are particularly susceptible. As yet, there are few reliable sets of experimental data available to quantify gaseous losses and they are generally not taken into account when making fertiliser recommendations.

AMMONIUM FIXATION ON EXCHANGE SURFACES

Some ammonium is held in the soil on clay mineral and organic matter surfaces. As the cation is readily nitrified to nitrate, the amount held in this form is small. Some ammonium is held by fixation in clay mineral lattices, but this is not exchangeable with other cations in solution. The amount held is usually small and is not a major mechanism of nitrogen fixation in UK soils.

NITROGEN FIXATION

Fixation of atmospheric nitrogen as ammonia can be carried out by a number of micro-organisms. There are two types of nitrogen fixation; one carried out by free-living micro-organisms in the soil and the other by micro-organisms living in symbiosis with plants. The biochemical mechanism for each is probably the same, both involving the enzyme nitrogenase. The quantity of nitrogen fixed by free-living micro-organisms is small under UK conditions. While being very difficult to measure in the field, annual amounts are probably no more than 5–10 kg/ha N per year, being lower under arable than under grassland.

Agriculturally, symbiotic fixation is far more important. Legumes and a few other species have the ability to fix atmospheric nitrogen. The legumes carry out the fixation in symbiosis with the soil bacteria, *Rhizobium*, which takes place in nodules located on the plant roots (see Plate 4.1). Nodule formation follows infection with the appropriate strain of *Rhizobium*. In some situations inoculation to introduce the bacteria may be necessary if a legume is to be grown on a field for the first time. The effectiveness of the symbiosis depends on environmental conditions. Fixation is inhibited below pH 6 and by high nitrogen fertiliser use. Well-aerated soil conditions are needed for good nodulation.

It is often assumed that as a result of growing a legume crop, considerable nitrogen is left in the soil. Most of the nitrogen fixed in the nodules is translocated to the above ground parts of the plant. Therefore the return to the soil depends on whether the whole plant is removed or some crop residue is left on the field. Perennial legumes such as lucerne or clover build up the soil nitrogen content considerably even when cut and removed, but the contribution from peas and beans is small if the haulm is removed or burnt. When examined, active nodules are pink inside. If this colour is absent the nodules are usually inactive. Nodule activity is often stopped by drought.

The amount of nitrogen per hectare fixed by a legume crop varies from 50 to 200 kg/ha for annual pulses up to 300 kg/ha for lucerne. Of more practical importance for fertiliser recommendations is the quantity remaining in the soil for use by any following non-leguminous crops. Clovers and lucerne may commonly leave 150–200 kg/ha nitrogen under temperate conditions. The amount left by short growing season crops of peas and beans is probably of the order 20–50 kg/ha nitrogen.

Not all legumes can be economically produced in the UK without fertiliser nitrogen. Both runner and dwarf beans respond to fertiliser nitrogen. Although they produce nodules, the amount of nitrogen fixed by these *Phaseolus* beans is much less than is fixed by *Vicia faba* (field and broad beans).

PREDICTION OF CROP FERTILISER REQUIREMENT

Our current ability to predict the nitrogen requirement of a crop is better than ten years ago but is still only approximate. The wide range of crops, soil types and other factors ensure that precision beyond the nearest 20 kg/ha N is unlikely to be achieved. Current efforts at crop modelling backed up by more sophisticated techniques of measuring nitrogen contributions from different sources provide hope for further improvements over the next few years.

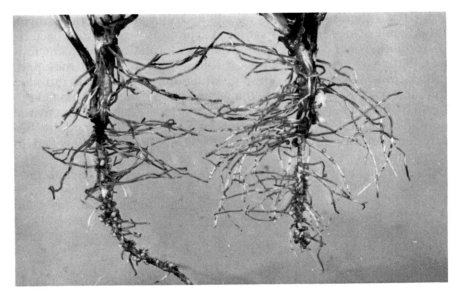

Plate 4.1. Nodulation of field beans

To give a precise recommendation, three factors must be known:

1. Crop demand.
2. Soil nitrogen supply.
3. Efficiency of fertiliser nitrogen uptake.

The main concern of this chapter is the prediction of the amount of soil nitrogen from the various processes discussed that will be taken up by the plant. One seemingly sensible approach has been to measure the mineral nitrogen (ammonium and nitrate) in the soil profile to 1 metre depth at the beginning of spring. This technique has shown some success when restricted to one soil type. While it is not surprising that a single factor fails to account for all the variation in experimental results, the combination of several factors demands a good under-standing of their relative contributions and how they interact. Profile mineral nitrogen measurements are most likely to be worth while in high nitrogen residue situations.

For several decades, researchers have made use of various laboratory incubation techniques to try to measure the amount of organic nitrogen likely to become available for crop uptake. Numerous individual techniques have followed the same general theme. This involves measuring the field soil mineral nitrogen content, incubating the soil

Plate 4.2. Lysimeter installation

Letcombe Laboratory

under specified conditions in the laboratory and then measuring the mineral nitrogen content. The hope is that the increase in mineral nitrogen content or net mineralisation that has taken place can then be related to soil nitrogen supply in the field. The testing is tedious and success has been limited to a narrow range of conditions.

EXPERIMENTAL TECHNIQUES

A number of techniques are employed in nitrogen research, often with the aim of trying to construct a balance sheet for a particular nitrogen cycle. It is all too easy to make the balance of losses and gains come to 100 per cent by assuming that the amount unaccounted for was lost to the atmosphere by denitrification or was leached. Recent work has attempted to measure all changes so that no assumptions have to be made.

Leaching studies are often carried out in lysimeters, which are enclosed volumes of soil, varying in size from less than one to several cubic metres in volume (see Plate 4.2). They may be filled with disturbed soil or carefully taken undisturbed cores. Their common feature is a closed base so that all drainage water can be collected.

Catchment studies attempt many of the same objectives as lysimeters but over areas of several hectares. The common approach is to find a field or valley on impervious clay where all the drainage water leaves the catchment by one ditch or stream. This is then flow monitored and the water analysed.

Another advance in the last few years has been the use of a heavy isotope of nitrogen, N–15, which has a slightly different atomic mass from the normal N–14. By labelling fertiliser or organic material with N–15, aspects of nitrogen use can be measured. A specific use has been the labelling of fertiliser applied to crops. By analysing the ratio of N–15 to N–14 in the crop, the relative contributions of fertiliser N and soil N to total crop uptake can be calculated.

Now that the technique exists to separate these two components, it is possible to examine in detail the various factors of soil type, winter weather and previous cropping which are known to influence the amount of nitrogen available for crop uptake during the growing season. Rather than ranking situations as low, medium or high as is currently done by the ADAS Nitrogen Index system, actual quantitative estimates of the kg/ha nitrogen available from a crop residue are possible. When this is added to a figure for the efficiency with which the fertiliser nitrogen is utilised, also from N–15 studies, a much more precise field specific recommendation can be derived.

Chapter 5

PHOSPHORUS

PHOSPHORUS IS taken up by plants in smaller amounts than nitrogen or potassium. Soil phosphorus levels have been built up in UK fields over the last fifty years; thus now fertiliser applications are often applied to maintain soil reserves rather than to give spectacular differences in growth. There are still many fields, especially in grassland areas, which will give crop yield responses to annual applications of phosphate fertiliser.

Phosphorus has attracted a tremendous amount of research effort over the last 100 years, but its behaviour in the soil and availability to crops are still imperfectly understood. One reason is the numerous inorganic and organic forms of phosphorus that occur in soils and the wide variation in behaviour between soil types. There are also major differences between crops in their ability to take up different forms of phosphate.

PLANT METABOLISM

The form in which phosphorus is available in the soil to be taken up by the plant root system depends primarily on the soil pH. The following equilibrium determines which orthophosphate anion is dominant in the soil solution.

$$HPO_4^{2-} + H^+ \rightleftharpoons H_2PO_4^-$$
$$\text{high pH} \qquad\qquad \text{low pH}$$

Generally the concentration of phosphorus in the plant xylem is 100 to 1,000 times more than in the external soil solution, indicating an active energy demanding uptake process. Once inside the plant, the phosphorus anions are rapidly metabolised into organic compounds. These move freely in both the xylem and phloem transport systems. Movement can also occur in inorganic form.

Phosphorus occurs in plants in numerous forms, such as phosphory-lated sugars and alcohols and as phospholipids which have a wide range of functions. The main function is in energy transfer as adenosine triphosphate (ATP). This contains high energy pyrophosphate bonds which are formed during photosynthesis and which, on hydrolysis, release this energy to drive other plant processes, including active uptake of plant nutrients.

ORGANIC PHOSPHORUS IN SOILS

While organic forms of phosphorus are an essential part of the soil organic matter, for most UK situations, organic phosphorus is far less important than inorganic forms in providing crop requirements (Figure 5.1).

Micro-organisms have a major requirement for phosphorus and their metabolism may be restricted by a lack of soil phosphorus. In the UK this is only likely to occur in unfertilised situations. When organic material is incorporated into the soil, its phosphorus content will not usually be low enough to limit microbial breakdown. Compared to a C:N ratio of 10 to 20, a C:P ratio of 100 to 200 is adequate for breakdown without the need for an external phosphorus source. The

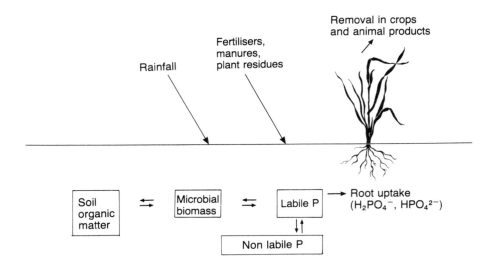

Figure 5.1. Phosphorus cycle

main form of organic phosphorus in the soil is inositol phosphate. This has to be broken down by enzyme action to inorganic orthophosphate before it can be taken up by the plant.

INORGANIC PHOSPHORUS IN SOILS

Numerous inorganic phosphorus compounds occur in soils. Phosphorus anions are held with varying strengths on numerous soil particle surfaces. The result is a complex situation compared to that for most other nutrients. Several calcium phosphates occur in soils. These have very different solubilities in the soil solution, but in general are only very slightly soluble. Phosphate fertilisers are derived from rock phosphate which is fluorapatite, a calcium-containing phosphate. Details of phosphate fertiliser materials are discussed in Chapter 9. Whilst some of the soil phosphorus-containing minerals occur in forms which can equilibrate with the soil solution, others are contained inside chalk or other nodules, unavailable for crop use.

To summarise a complicated system, the inorganic soil phosphorus that is relevant to crop uptake is held tightly in the soil in association with calcium, aluminium and ferric iron. These may occur as direct phosphate compounds, or in association with hydroxides, particularly of aluminium and iron. Much is also held on the broken edges of clay minerals particularly in association with aluminium. Some is adsorbed on calcium carbonate surfaces. Unlike the other anionic nutrients, the phosphate anions are strongly and rapidly adsorbed in almost all soil types.

PHOSPHORUS DEFICIENCY

Cases of visual symptoms of phosphorus deficiency associated with extremely low soil phosphorus levels are rare in the UK due to the general use of phosphate fertilisers. Most deficient crops show an overall stunting in early growth where phosphorus uptake is inadequate. This can be quite spectacular in winter wheat or oilseed rape but does not necessarily incur a reduction in yield.

Many spring-sown crops will exhibit classical symptoms of reddening or purpling of older leaves and stems due to increased anthocyanin production. A low soil phosphorus index in a cold spring may induce symptoms in spring barley. The UK crop most likely to show these symptoms is maize (see Plate C.4). Sometimes this is induced by poor soil conditions and low temperature, when the soil phosphorus level is satisfactory. Leaf levels of phosphorus in plants are commonly 0.3 to 0.4 per cent, while severe deficiency may give rise to less than half this value. Deficiency may produce stunted growth, but with a near normal leaf phosphorus concentration. Seeds and grains have relatively high phosphorus levels, while that of straw is low.

Foliar applications of phosphorus may be recommended as a remedial treatment in exceptional circumstances. For details see Chapter 14. Foliar application is used to increase the phosphorus content of some apple varieties to improve storage quality.

BEHAVIOUR IN SOIL

Due to its strong specific adsorption in soils, phosphorus behaviour is dominated by very low concentrations in the soil solution. This results in relatively poor mobility of phosphorus in soils, compared to cations or non-absorbed anions. The very low concentration of orthophosphate maintained in the soil solution (often less than 1 mg/l phosphorus) coupled with the considerable crop demand means that diffusion of ions to the root surface is very important in determining uptake.

The main source of plant-available phosphorus is generally termed the labile pool. This provides fairly rapid exchange with the soil solution, maintaining the solution concentration. The remaining fraction is the non-labile pool. This contains the large quantity of insoluble phosphate which is only very slowly released into the labile pool. The various organic and inorganic phosphates constitute these labile and non-labile pools. There is no clear distinction by which particular forms can be designated as in one or other pool. In general the labile pool can be considered as orthophosphate adsorbed on the surfaces of clay minerals, hydrous oxides and carbonates plus iron and aluminium phosphates. The relationship between the quantity of phosphorus in the labile pool and the soil solution concentration depends particularly on soil texture and pH. Adsorption of ortho-phosphate is stronger and therefore phosphate is less available at low pH. In calcareous soils, precipitation of calcium phosphates is a major factor limiting solubility. Phosphate availability is generally greatest between pH 6.0 and 7.0.

Loss of phosphorus to the non-labile pool occurs mainly by the formation of slightly soluble compounds. Phosphate adsorbed on calcium carbonate will slowly convert to the mineral, apatite. Precipitation as iron and aluminium phosphates will also reduce phosphorus availability. Fixation as insoluble compounds is most likely in calcareous soils.

Current evidence would suggest that on most soils the majority of the non-labile pool is not completely lost from plant uptake. Although in only slightly soluble form, it does contribute to the soil solution concentration particularly under conditions of intense depletion due to crop uptake. The practical consequences for fertiliser recommendations are discussed in Chapter 13.

Whatever happens to phosphorus in the soil, it is not lost by leaching. High amounts of inorganic or organic fertilisers applied to agricultural soils in UK are unlikely to lead to leaching losses greater than 1 kg/ha P_2O_5 per year. The main losses of soil phosphorus from the land occur due to wind or water erosion.

CROP UPTAKE

Crop uptake of phosphate anions depends predominantly on diffusion. The rate of diffusion of orthophosphate is typically 100 times less than that of potassium ions. Mass flow provides very little phosphorus due to the low soil solution concentrations. Therefore phosphate uptake is more dependent on aspects of crop root activity than is the case for other major nutrients.

This is emphasised by the fact that most studies show only 5–10 per cent of a fresh phosphate fertiliser application is taken up by the first crop. The figure will be higher in percentage terms if the soil phosphorus level is very low. The remainder is supplied by the soil reserves. Without physical mixing, the movement of fertiliser phosphorus in soil is measured in millimetres (mm) rather than centimetres (cm).

One of the main crop factors influencing phosphorus uptake is the extent of the root system in the enriched topsoil layer. The more highly branched the root system, the greater the volume of soil that will be within diffusion range. Root hairs have been shown to be particularly important in extending the volume of soil explored and increasing the amount of phosphorus taken up. Diffusion is strongly temperature dependent, so uptake in cold springs will be less than under comparable conditions at a higher soil temperature.

The rate at which the crop demands phosphate, particularly in early growth before the root system has explored a large volume of soil, is also a factor. Spring-sown crops are more likely to benefit from high levels of soil phosphorus than autumn-sown or perennial crops.

Species vary in ability to take up their phosphorus requirement. Potatoes require high concentrations of water-soluble phosphorus fertiliser or a very high soil phosphorus level for maximum yield, while grass and cereals can take up enough phosphorus despite a low soil phosphorus level. The importance of root association with particular micro-organisms known as endotropic mycorrhizae has been demonstrated to increase phosphorus uptake by a number of crops. Mycorrhizae occur naturally on some crops but not on others. At present it seems that this will only be of agronomic significance in the UK on very low phosphorus soils.

The influence of chemical and microbial activity in the rhizosphere is important. This is the zone around the root influenced by root exudates. Differences in pH and increased biological activity within the rhizosphere can both lead to considerably increased phosphorus uptake. Again this will be more beneficial if soil phosphorus supply is marginal.

Whatever the variation between crops, the variation in soil phosphate supply will depend on the concentration of phosphorus in the soil solution and the buffer capacity or ability to replenish that solution from the labile pool. If the buffer capacity is high, a lower soil solution concentration may be adequate. Crops need an adequate phosphorus supply throughout their growing season, but particular importance has been attributed to adequate phosphate for young plants. This is probably because uptake is most likely to be limiting at this stage. If the soil provides an adequate early supply it is likely to be able to sustain growth throughout the season. Placement of water-soluble fertiliser in close proximity to the seed frequently gives improved growth and yields for a number of species on low phosphorus soils.

RESIDUAL VALUE OF PHOSPHATE FERTILISERS

The first year recovery of applied fertiliser in a crop is generally below 15 per cent. Often the recovery over the three years following application is only 20–25 per cent of the original dressing. Very long term studies at Rothamsted show that after many decades without phosphate fertiliser, crops are still taking up 1 per cent per year of phosphate originally applied. This adds weight to the view that once the soil's initial fixation capacity has been satisfied, most of the phosphate subsequently applied is not irretrievably lost for crop use. Present evidence suggests that there are few if any soils in the UK which continue to adsorb phosphate fertiliser, so preventing the build-up of soil analysis levels.

A feature of crop response to phosphate is the failure to achieve maximum yield from an annual application on soils of low phosphorus reserves. Figure 5.2 shows this for potatoes at Rothamsted. On the low phosphorus soil, the yield response continues up to the highest rate of

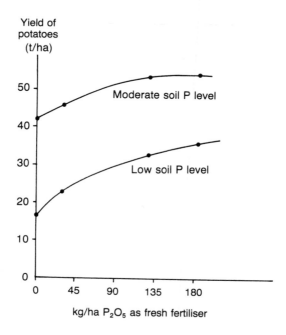

Figure 5.2. Response of potatoes to fresh and residual phosphate fertiliser
Source: Rothamsted Experimental Station

fresh fertiliser tested. In contrast, the high residue site gives a higher yield but a lower optimal level of fresh phosphorus fertiliser.

PRACTICAL ASPECTS

In predominantly grassland areas, particularly in the uplands of the UK, soil phosphate levels are often low and regular fertiliser phosphorus is needed to ensure maximum yields. In contrast previous fertiliser use on much arable land in the UK has raised levels so that only a few crops show yield responses. Many fields are at a soil phosphate level which does not justify further applications, except for potatoes.

Olsen's sodium bicarbonate method of soil analysis is most commonly used in the UK. It is very satisfactory on soils at pH 6.0 and above but is less satisfactory on acid soils. Of more concern is its inability to reflect the use of insoluble rock phosphates on grassland. An acid extraction method is more likely to prove worthwhile for these situations. In recent years the main field experimental effort has been directed towards trying to resolve the relative effectiveness and limitations of the numerous insoluble phosphate fertilisers that have come onto the UK market, following the termination of UK basic slag production.

Chapter 6

POTASSIUM AND SODIUM

THE MAJOR soil cations are calcium, magnesium, potassium and sodium. Of these the behaviour of potassium is unique in being preferentially held by some clay minerals. This is in contrast to the monovalent cation, sodium, which is weakly adsorbed and is readily leached. The behaviour of potassium in the soil is strongly influenced by soil texture, specifically clay content and mineralogy. Neither cation exists in organic forms. Potassium is second only to nitrogen in terms of the quantity needed by crops. Sodium is not an essential element for plants, but is a major nutrient in animal nutrition. Although not essential for crops some, such as sugar beet, show yield responses to the application of sodium fertiliser.

POTASSIUM

Plant Metabolism

Unlike phosphorus, potassium occurs in the soil solution as the simple cation K^+. The high uptake rates of K^+ necessary to satisfy crop demand are achieved by a specific, energy using, active process. Once in the plant, potassium is extremely mobile in the xylem and phloem. It remains almost exclusively in ionic form in the cell sap and is easily redistributed within the plant. While there is competition between cations for plant uptake from the soil solution, potassium generally has a dominant influence. Much of the crop uptake of potassium occurs during the vegetative stage of crop growth; in cereals uptake is largely completed by ear emergence.

The main role of ionic potassium in the plant is to regulate the cell water content. Maintaining cell turgor and maximum rates of cell expansion depends on cell potassium levels. Transpiration rate also depends on the presence of adequate potassium in the guard cells controlling stomatal movement of the leaves. Most plant processes, particularly the translocation of photosynthates within the plant, are dependent on cell potassium concentrations. Potassium is also essential to a number of plant enzyme systems. Unlike other essential nutrients there are few if any plant processes in which potassium forms part of an organic molecule. Its role is to regulate processes rather than to become fixed in one form.

Potassium Deficiency
In the UK, the extreme of visual symptoms of potassium deficiency are occasionally seen on crops grown on sandy or chalk soils. More common, but much less easily identified, is a reduction in plant size due to inadequate potassium uptake. Plants generally show a loss of turgor, making them less drought resistant. Wilting occurs under high evaporation conditions and crops are more susceptible to frost damage. Visual symptoms of extreme deficiency are invariably seen first on the older leaves. Most species initially show a marginal chlorosis, spreading towards the main vein as the deficiency becomes worse. Cereals show a characteristic bleaching or chlorosis of the leaf tip which progresses down the leaf as symptoms intensify. Clover shows interveinal chlorotic and necrotic spots. In some crops the occurrence of marginal chlorosis separates potassium from magnesium deficiency symptoms very easily, eg. in potato (see Plate C.5) and apple, while in others such as oilseed rape the separation is much more difficult as both give marginal symptoms. Here the colour may be a useful guide; magnesium deficiency symptoms generally show more anthrocyanin purpling.

Soil Potassium
The behaviour of potassium in the soil and the availability for crop uptake depends on the general relationship below:

$$\text{native } K^+ \rightarrow \text{non exchangeable } K^+ \underset{\text{slow}}{\rightleftharpoons} \text{exchangeable } K^+ \underset{\text{fast}}{\rightleftharpoons} \text{soil solution } K^+$$

Most soils contain native potassium which is not readily available for crop uptake. Depending on clay content, soils may contain up to 3 or 4 per cent of total potassium. The key aspect of potassium behaviour in a particular soil is its ability to maintain an adequate soil solution level during rapid crop growth and potassium uptake. This depends on the soil holding enough potassium in exchangeable form. Once any initial fixation capacity has been satisfied, there is generally a close relationship between exchangeable potassium and the ability of the soil to provide a satisfactory solution concentration. Of longer term importance is the capacity of the soil to replenish the exchangeable supply. This may come entirely from non-exchangeable forms held in the clay minerals or from fertiliser addition.

Clay Mineralogy

The feldspar group of minerals contain large amounts of potassium which is constantly being released as the minerals weather and break down into smaller units. These provide a slow release of native potassium. Various clay minerals occur in soils; all hold cations in exchangeable form on their surfaces. They vary considerably in their capacity to hold potassium in non-exchangeable forms. As shown in Figure 6.1 there are two basic types. The 1:1 lattice types such as kaolinite do not have interlattice spaces, while the 2:1 types have this property. The ability of the latter to hold potassium preferentially between the structural lattices provides the source of non-exchangeable potassium.

Most geologically old soils in the UK contain predominantly kaolinitic clay minerals. These are non-expanded 1:1 types which hold no non-exchangeable potassium trapped between layers of clay mineral. They have a low CEC and behave as sand or organic matter in their potassium relations. At the other extreme illites, vermiculite and chlorite selectively adsorb large amounts of potassium which is fixed between the lattice layers and slowly exchanged with the exchangeable form. Montmorillonite holds an intermediate amount in non-exchange-

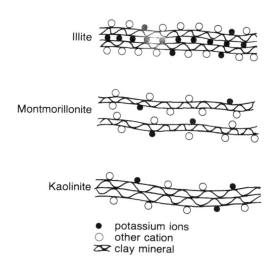

Illite

Montmorillonite

Kaolinite

●   potassium ions
○   other cation
⋜⋝   clay mineral

Figure 6.1. Structure of clay minerals

able form. As these various minerals weather, considerable potassium is released.

Most clays in UK soils contain a mixture of clay minerals. While soils high in 2:1 clay minerals generally provide a large reserve of potassium, depletion of the reserve by exhaustive cropping, eg. cut grass receiving no potassium fertiliser, will result in a high rate of fixation of any subsequent potassium fertiliser applied. Fixation will compete favourably with crop uptake for potassium under these conditions. Adequate potassium must be applied or released by weathering to maintain saturation of the 2:1 minerals. Otherwise the non-exchangeable potassium level will fall. Weathering is often adequate to achieve this except in high offtake situations.

The level of exchangeable potassium needed to maintain both the soil solution level and saturation of the non-exchangeable sites depends on the overall clay CEC. Sandy soils need a lower exchangeable level than most clays to give the same levels of supply to the soil solution. The major difference in potassium supply due to clay content is the ability of the different textures to maintain the exchangeable levels needed. On sands, regular fertiliser input is needed whereas on clays the higher exchangeable level is often adequately maintained by weathering of clay minerals.

Soils with a moderate 2:1 clay mineral content lose little potassium by leaching. The normal annual loss is only 1–2 kg/ha K_2O. Only soils with virtually no expanded lattice clay content lose appreciable amounts of potassium by leaching. Even 5 per cent of clay will reduce the leaching risk to a low level, as long as application rates are not much above those removed in crops.

Crop Uptake

While roots take up more potassium than phosphorus by mass flow, diffusion is the major mechanism by which the potassium requirement of the crop is satisfied. The soil solution concentration in the root zone is rapidly depleted during periods of rapid growth and constant diffusion from the exchangeable supply provides for this demand. The rate of diffusion will only be adequate if the expanded clay minerals are potassium saturated. This will usually be the case for UK soils where potassium fertilisers have been applied. Most soil analysis methods for available potassium measure the exchangeable plus solution content.

Diffusion rates are lower in dry soils compared to soil at field capacity. This is probably the reason for deficiency symptoms sometimes seen in potato crops in dry seasons, particularly on sandy soils. In this situation a high fertiliser application rate is not always enough to give maximum growth.

Field Aspects

Soil analysis for potassium and its use in predicting fertiliser requirement are discussed in Chapter 12. Soil solution concentrations and potash reserve measurements including the non-exchangeable forms may be used to help explain experimental results or advisory problems where the normal exchangeable measurement gives an inadequate explanation.

The level of available potassium that can be maintained in a soil depends mainly on its clay content. Soils with above 20 per cent of mainly expanded lattice clays can retain very high levels, up to 300–400 mg/l K. In contrast, levels above 100 mg/l cannot be maintained on loamy sands with only 5 per cent clay. On the latter soils the only practical method of building up soil available K levels is by using frequent heavy applications of farmyard manure. By this method levels of 200 mg/l can be maintained, due to an increased CEC from the organic matter.

On soils of high available potassium, most vegetative crops will take up potassium in proportion to the soil analysis level. This capacity to take up luxury amounts of potassium is sometimes a practical problem. Firstly magnesium uptake may be reduced resulting in crop symptoms and yield loss or animal nutrition problems. Secondly it is very wasteful if the potash is removed from the field, as occurs in dried grass production and is not returned in slurry or farmyard manure.

SODIUM

Plant Metabolism

While sodium is essential for some algae, the element is not essential for higher plants. Some UK crop plants need sodium for maximum yield. A considerable part of the osmotic regulatory function in these plants can be carried out by either sodium or potassium ions, but for sugar beet at least, both are needed for best results. Each can replace the other to some extent.

Sodium increases the drought resistance of plants; it reduces the extent of wilting and increases the rate of leaf expansion. Much of the literature on sodium relates to problems of excess sodium in arid climates or following sea flooding. These aspects are outside the scope of this book.

Uptake of sodium from the soil solution is in straightforward cationic form as Na^+. Mass flow is likely to provide the majority of the crop uptake under most circumstances. The quantity taken up varies considerably with different crops indicating that some uptake selectivity occurs between species. The crops that show a yield response to sodium are generally of maritime origin. The main UK crops falling in this category are shown in Table 6.1. Other crops may respond under low soil potassium situations. No crop shows readily recognisable symptoms of sodium deficiency.

Table 6.1. Crops responding to sodium

Responsive when K adequate	Responsive when K deficient
Sugar beet	Barley
Fodder beet	Wheat
Redbeet	Oats
Mangolds	Peas
Turnips	Cabbage
Celery	Kale
Carrots	

Soil Sodium

The majority of sodium in the soil is either in solution or on the clay surfaces in exchangeable form. For most soils the exchangeable sodium content is very low compared to calcium or magnesium and generally lower than for potassium.

Sodium is not preferentially adsorbed in non-exchangeable forms. A small amount is released annually in clay soils by weathering, but in the absence of fertiliser sodium, the soil exchangeable level is controlled by rainfall inputs. These vary from 10 to 20 kg/ha Na in inland sites up to 50–100 kg/ha Na on western coastal sites.

Leaching of sodium occurs on all soils if the element is not removed in the crop. The speed and efficiency of removal follow the general principles of leaching discussed in Chapter 2. Sodium is relatively weakly held on exchange sites, compared to divalent cations. Some studies have shown up to 50 per cent of the soil sodium present in the soil solution. It is more susceptible to leaching than any other cation. For this reason it is not possible to build up soil sodium levels over a period of years. Winter leaching removes a considerable proportion of an annual application. Most responsive crops are only grown in rotation one in three or one in four years and therefore need fertiliser sodium each time they are grown.

The levels of available sodium in sugar beet growing soils has been the subject of a number of studies. Each has shown that soil level varies with soil texture. The main variation is between organic and mineral soils. Nearly all mineral soils are below 20 mg/l Na as shown in Table 6.2. By contrast most organic soils are above 20 and many are above 40 mg/l Na. The high percentage of low sodium, sandy soils reflects low weathering contributions and low exchangeable levels, coupled with efficient leaching of rainfall inputs.

Table 6.2. Typical exchangeable sodium levels in East Anglian sugar beet soils

	mg/l Na
Sand	15
Light loam	18
Medium loam	24
Silt	30
Clay	30
Organic	50

Source: ADAS.

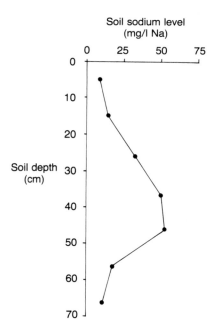

Figure 6.2. Movement of sodium by leaching on sandy loam
Source: Brooms Barn Experimental Station

As part of the same study, the movement over winter of autumn-applied sodium down soil profiles of various textures was examined. Autumn-applied sodium at 800 kg/ha NaCl (300 kg/ha Na) was followed by autumn ploughing and soil sampling to 1 m depth in 10 cm increments in March. The graph (Figure 6.2) shows the movement that had occurred. Excess winter rainfall was 160–200 mm, typical quantities for East Anglia. The study also showed that the subsoil sodium content of marine silt soils is commonly high compared to other soils and increases with depth and proximity to the saline groundwater.

Field Aspects
A major input of sodium to soils can be via irrigation water. Water sources vary from 50 to 1,000 mg/l Na. In the latter situations, soil levels of sodium for responsive crops even on sandy soils are often adequate if irrigation has been applied in the previous year. Similarly fields very close to the sea will often receive adequate sodium in rainfall.

A high level of exchangeable sodium in soils can disperse clay particles, resulting in instability problems. This is most dramatically seen following the sea flooding of arable silts and clays. For this reason, farmers are often concerned at the possible adverse effects of applying fertiliser sodium to these soil textures. A series of ADAS experiments have studied this on unstable silty soils. At rates of application several times those recommended for responsive crops, no deleterious effects could be demonstrated. The amount of salt applied to sea-flooded land is many times that recommended for sugar beet.

Chapter 7

MAGNESIUM AND SULPHUR

THE TWO remaining major nutrients have few similarities but for convenience are dealt with in this chapter. As shown in Table 2.2, they are required in similar quantities by crops.

Magnesium is adequately supplied by soils with moderate or high clay contents. Deficiency and the need to use magnesium fertilisers is generally restricted to sandy soils. There is variation in crop requirement. Not all crops on low magnesium soils will respond to magnesium fertiliser. Sugar beet have attracted the major share of the experimental work done in the UK.

Sulphur and the need for sulphur fertilisers have attracted considerable discussion in recent years. Current information suggests that for England and Wales atmospheric inputs of sulphur are adequate for most arable crop requirements. Deficiency occurs in grass cut for silage in many parts of northern and western Britain.

MAGNESIUM

Plant Metabolism

For most situations, mass flow provides the major proportion of crop magnesium needs. Uptake is as the cation Mg^{2+}. The amount of magnesium taken up is influenced by the quantity of other cations in the root environment. Numerous experiments have shown a strong antagonism with potassium ions. High soil levels of potassium reduce the uptake of magnesium and can intensify deficiency problems. Table 7.1 shows a typical example of the interaction of these two nutrients as shown in leaf analysis levels. Ammonium ions also antagonise magnesium uptake.

Magnesium is highly mobile in the phloem as well as in the xylem and is freely translocated around the plant. For this reason older leaves will generally be the first to show deficiency symptoms. Whereas leaf

Table 7.1. Effect of potassium fertiliser application on magnesium uptake by grass

	K₂O applied kg/ha		
	28	56	84
	Magnesium uptake % Mg (DM)		
Autumn applied	0.174	0.169	0.163
Spring applied	0.170	0.162	0.152

Source: West of Scotland Agricultural College.

levels of magnesium are inversely related to potassium levels, fruits and storage organs, notably potatoes and apples, show the opposite effect. For these parts of the plant, it is usual to find magnesium and potassium levels increasing in direct relationship.

The best known biochemical role of magnesium in plant tissue is its position in the chlorophyll molecule. Chlorophyll is the main pigment involved in the photosynthesis process. However, no more than 25 per cent of the magnesium in the plant performs this function. Other functions include involvement in a large number of enzyme systems.

Magnesium Deficiency
For most species, visual symptoms are described although they may sometimes be confused with other nutrient deficiencies. Deficiency is first seen on the older leaves or on the lower leaves of annual extension growth in top fruit (see Table 7.2). While the detailed symptoms vary between species, most broadleafed species show interveinal chlorosis, with a broad veinal area remaining green. Some species show marginal chlorosis as well, which can be difficult to distinguish from potassium deficiency. In addition brassicas and some other species can show intense anthrocyanin purpling. In severe deficiencies the chlorotic areas turn necrotic and may drop out leaving holes in the leaf.

The general symptom expression is caused by disruption of chloro-plasts and loss of chlorophyll, resulting in the initial leaf chlorosis. The level of leaf magnesium associated with deficiency is a very useful confirmation of visual diagnosis but threshold levels vary considerably with species.

Soil Magnesium
Soils contain magnesium in several mineral constituents including a number of clay minerals. The natural weathering of these minerals is more than adequate to provide the magnesium requirement of crops

Table 7.2. Symptoms of magnesium deficiency

Cereals	Green beading occurs in a regular pattern between the veins against a yellow/green background in the older leaves. The symptoms are best observed by holding a leaf up to the light. Interveinal and marginal necrosis may occur in severe cases.
Brassicas	Symptoms are normally seen in the older leaves. A marbling pattern of yellow and sometimes orange/purple colours develops between the veins from the leaf margins (see Plate C.8). Deficiency is often induced by drought and associated with nitrogen deficiency symptoms.
Sugar beet	Symptoms first appear in older leaves but may progress to the younger ones. Initially leaves show interveinal chlorosis near the margins which extends to midrib (see Plate C.7). These chlorotic areas become necrotic and holes may appear in the leaf.
Potatoes	Older leaves are commonly affected during senescence. A bright yellow chlorosis followed by necrosis occurs between the veins, while the leaf margin remains green (see Plate C.6).
Dwarf beans	Interveinal yellowing of older leaves occurs with leaf margins remaining green. Necrosis often follows.
Carrots	Older leaves become chlorotic and leaf margins show yellow/orange colours. It is difficult to distinguish between magnesium and nitrogen deficiencies.
Apples	Interveinal necrosis occurs in most varieties while the leaf veins and margins remain green. Premature leaf drop from the lower third of the extension growth commonly follows.

on all except very sandy soils, which have a very low CEC. As for potassium, magnesium exists in soil solution and in both exchangeable and non-exchangeable forms. The rate of conversion of non-exchangeable to exchangeable is slow and only slowly replenishes the exchangeable form taken up by crops. The concentration in the soil solution is generally relatively high, hence mass flow is the predominant transport mechanism to the root surface. Very little magnesium is associated with the soil organic matter.

Like calcium, magnesium is leached relatively easily from the soil. It is not preferentially adsorbed by clay minerals like potassium. The mechanisms of magnesium loss are similar to those for calcium and depend in part on anion additions in fertilisers. As with calcium, losses of magnesium will be increased if large amounts of non-absorbed anions such as chloride and nitrate are leached. The balance of exchangeable magnesium remaining in the soil after winter leaching also depends on the rate of magnesium release from soil mineral weathering. It is only necessary to apply magnesium fertilisers on sandy soils of low cation exchange capacity to maintain an adequate level of magnesium in the soil to meet crop demand. In these situations, crop offtake plus leaching is greater than release by weathering and depletion of soil levels will occur if fertiliser is not used. Rainfall inputs are generally around 5 kg/ha Mg per year away from coastal sites. Coastal sites receive around 10 kg/ha Mg per year which is a major contribution to maintaining soil levels.

Some soil parent materials contain appreciable magnesium and result in very high available magnesium levels in soil. These levels are unlikely to have an adverse effect on uptake of other nutrients. Particular soils in this category are those derived from magnesian limestone, Keuper Marl and marine deposits. Sandy soils on which the traditional liming material is magnesian limestone have much greater reserves of available magnesium than those limed with ordinary chalk or limestone. The Bunter sandstone soils of Nottinghamshire which occur adjacent to the magnesian limestone formation have much higher soil magnesium levels than similar textured soils in East Anglia. In such situations where the liming material contains magnesium carbonate, the magnesium level in the soil will be maintained at an adequate level for crop production, as long as the pH is maintained.

Crop Response
The amount of magnesium removed in crops is generally 10–20 kg/ha Mg per year. Cereals are at the lower end; root crops and brassicas at the upper end. The two main arable crops for which responses have been shown experimentally are sugar beet and potatoes. Forage and vegetable brassicas are also responsive, as are carrots. While no UK work has been carried out on oilseed rape, response at soil index 0 is likely.

Work by Brooms Barn Experimental Station on sugar beet (see Chapter 17) has clearly shown that soil analysis for available magnesium gives a good relationship with crop response. On soils with less than 25 mg/l available Mg, as measured by the ADAS method (see Chapter 12), yield response is probable for sugar beet and other responsive crops. Response is likely at higher soil magnesium levels for some fruit crops, and for potatoes receiving high rates of potassium, especially if the soil potassium level is also high. On most sandy soils, which are low in exchangeable magnesium, the exchangeable potassium is also low. Most induced deficiencies of magnesium result from excessive use of potassium fertilisers. Cereals do not respond to magnesium unless the soil level is very low—usually below 15 mg/1Mg.

Potassium – Magnesium Ratios

As mentioned earlier in the chapter, high potassium levels either in the soil or due to applied fertiliser generally cause a reduced uptake of magnesium. For this reason considerable emphasis has been put on K:Mg ratios in the past, when making fertiliser recommendations. There are few situations in field crops where the ratio is of concern. For annual crops, magnesium should be applied on the basis of soil analysis. This will be satisfactory whatever the soil potassium level, as long as the appropriate potash application is also applied.

The ratio is more important in fruit crops, but again the main problems arise where excessive potash levels have been applied over a few years resulting in very high soil potassium levels. Thus the need to consider the ratio is limited to low soil magnesium situations where the potash applied is or has been above the generally recommended rates.

Field Aspects

Magnesium deficiency symptoms are commonly seen in crops and confirmed by leaf analysis, even when the soil level of magnesium is adequate. In many crops, magnesium deficiency symptoms are commonly related to stress factors. The most frequent is the association of symptoms with poor soil physical conditions resulting in restricted root development. Symptoms are generally induced by a spell of either wet or dry weather which puts the poorly rooted crop under stress. Not all stress symptoms on crops are due to poor magnesium uptake. Drought alone can induce magnesium deficiency symptoms. These are common in potatoes often associated with premature senescence. Many crops show magnesium deficiency in the older leaves as senescence takes place.

Few if any of these induced deficiencies are likely to respond to foliar magnesium applications. The symptoms are often associated with nitrogen deficiency and response to extra nitrogen is more likely than to magnesium. Some crops, notably oilseed rape, can show transient magnesium deficiency symptoms on soils with adequate magnesium levels during periods of rapid growth and high nutrient demand. It is always the older leaves that are affected and treatment is unlikely to be worthwhile. Treatment by foliar applications (see Chapter 14) of epsom salts is likely to be worthwhile on low magnesium soils where deficiency symptoms are seen early in the life of a responsive crop. Otherwise yield response to foliar application is unlikely.

SULPHUR

Over the last few years, sulphur deficiency and crop response to sulphur fertiliser have become more important in the UK. Sulphur responses have been shown on arable crops and on grass in northern Scotland and in Ireland. Yield responses on second and third silage cuts are common in south-west England and Wales. Only small responses have been shown to date on arable crops in these areas. As atmospheric sulphur inputs are reduced further, greater and more widespread responses are anticipated.

Plant Metabolism

Sulphur is taken up by plant roots as the sulphate anion, SO_4^{2-}. Translocation is mainly by the xylem with little phloem movement. Plants can also utilise gaseous atmospheric sulphur, mainly sulphur dioxide SO_2. It is taken in through the leaves and distributed throughout the plant. Either soil or atmospheric sources can supply the major part of the crop requirement. High levels of atmospheric sulphur dioxide can damage plants.

Once in the plant, sulphur is reduced and combined in amino acids. Cysteine and methionine are the two most important sulphur-containing amino acids in plants. Sulphur is also an important component of the oil in oilseed rape.

Sulphur Deficiency

Sulphur deficiency occurs in several countries of the world. Chlorosis due to sulphur deficiency resembles nitrogen deficiency in many respects, except that symptoms generally appear in the younger leaves first, emphasising the poor mobility of sulphur within the plant. Oilseed rape in France shows a chlorotic marbling of the younger leaves during rapid spring growth. Legumes, particular lucerne, grass and brassicas are the UK crops with the greatest sulphur demand. Leaf analysis is a good indicator of crop sulphur status.

Soil Sulphur

Sulphur occurs in both inorganic and organic forms in the soil (Figure 7.1). Organic sulphur is the larger part of the total in most soils. Soil organic matter has a carbon:sulphur ratio (C:S ratio) of about 100:1. Like phosphorus, sulphur is a major nutrient needed by soil micro-organisms and a considerable amount of sulphur is constantly being cycled as organic matter is decomposed. Unlike phosphorus, the amount of inorganic sulphur in most soils is relatively small.

The inorganic sulphur provides the major part of the supply of

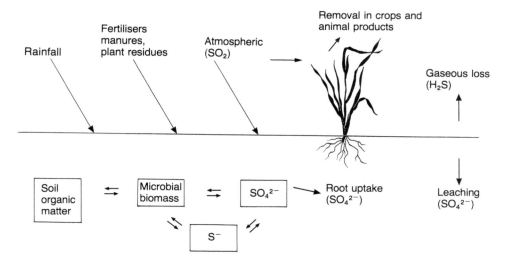

Figure 7.1. Sulphur cycle

sulphate taken up by the roots. Sulphate is present in soil solution and also adsorbed by some clay minerals and on the surfaces of iron and aluminium compounds. Sulphate compounds, particularly the slightly soluble gypsum, occur in some subsoils. Sulphate is held more strongly than nitrate or chloride but much less strongly than phosphate. It is fairly easily leached.

Soils most likely to show deficiencies in England and Wales are sandy, free-drained alkaline soils, low in organic matter and growing crops removing considerable quantities of sulphur. Deficiency is more probable after a wet winter. The other factor is a complete lack of sulphur applied in either inorganic or organic fertilisers.

Atmospheric Sulphur

Sulphur is deposited in rainfall and is absorbed by soils and crops direct from the atmosphere. These two inputs together apply considerable sulphur per hectare annually. The main source of atmospheric sulphur is from the burning of fossil fuels. While atmospheric pollution has been reduced in urban areas in recent years, the change has been much less in rural areas. The amount of sulphur deposited in an area depends on proximity to urban and industrial emissions. It also depends on rainfall and the dominant wind direction. The sea is a minor source of atmospheric sulphur, only influencing a few kilometres (km) from the

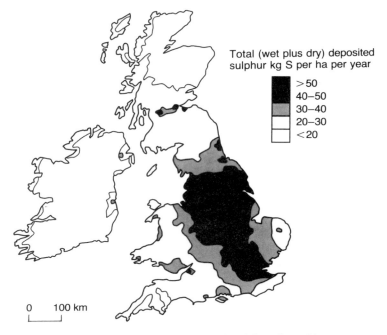

Total (wet plus dry) deposited
sulphur kg S per ha per year

■	>50
■	40–50
▨	30–40
□	20–30
□	<20

0   100 km

Figure 7.2. Total (wet and dry) sulphur deposition

coast. The amount of sulphur in rainfall also influences lime loss (see Chapter 3).

Rainfall contains sulphates and free acids, while the main form in the atmosphere is sulphur dioxide. The latter provides a major part of the crop sulphur requirement by direct absorption through the leaves. Recent work by the Central Electricity Generating Board has estimated the total amount of wet plus dry deposition of sulphur for the British Isles as illustrated in Figure 7.2. The main feature is the strong trend of increasing sulphur supply from west to east. Over most of England except the south west, annual input is 30 kg/ha S or above. The main areas of relatively low input are north Scotland and Ireland.

Crop Response

Field responses to sulphur fertilisers by arable crops are generally small in the UK. Oilseed rape has shown small but statistically significant yield increases in Dorset and in northern Scotland, both areas with 20 kg/ha S or less of annual sulphur deposition. Most cereal experiments have failed to show a sulphur response.

By contrast, multicut silage in Wales, south-west England, Scotland and Ireland has shown substantial yield responses. Large responses are generally confined to sandy and light chalk soils. The highest annual dry matter yield benefit shown in recent ADAS experiments was 47 per cent on a Dorset chalk site. Typical sites show no significant benefit at first cut followed by substantial second and third cut yield responses. Recent experiments in south-west England have shown that slurry can provide some of the sulphur fertiliser requirement of grass cut for silage.

Table 7.3. Sulphur-containing fertilisers

	Sulphur %
Gypsum	14–18
Ammonium sulphate	24
Superphosphate	11–14
Kieserite	23
Potassium sulphate	18

It is likely that more cases of response to sulphur will be shown in south-west England, Wales, north Scotland and Ireland. In most of these predominantly grassland areas, organic manures recycle sulphur and soil organic matter levels are generally higher than further east. Thus areas of deficiency are likely to remain localised. Work in New Zealand has shown that clover growth is depressed in mixed swards under sulphur-deficient conditions. The sulphur content of the crops varies, as shown in Table 2.3.

The main sulphur-containing fertilisers which might be applied to sulphur-deficient situations in the UK are given in Table 7.3. Where sulphur is applied as a foliar application to growing crops, any yield benefit achieved may be due to either nutrient benefit or disease control. Sulphur is a broad spectrum fungicide and yield response due to disease control is difficult to discount even in the presence of an apparently adequate fungicide programme. Only in marginal nutrient deficiency situations is foliar application likely to give the full sulphur response obtainable from soil application.

Soil and Plant Analysis
Due to the behaviour of sulphate in soils, soil analysis levels tend to be low with small variation. Much depends on the soil organic matter content. Choice of analytical method is generally not critical. Water

extraction is commonly used; potassium or calcium di-hydrogen phosphate solution has been used recently by the Macaulay Institute in Scotland and by ADAS. If an area becomes deficient due to low atmospheric deposition, soil type and rotation may provide an adequate basis for deciding on responsive situations, as is done for sodium. It is not feasible to build up soil sulphur levels by inorganic fertiliser use.

Plant analyses for N:S ratio and for total S content have both proved useful in assessing the likelihood of a sulphur fertiliser response. Both methods have the advantage of taking atmospheric sulphur uptake into account. Plant analysis has the major disadvantage of not enabling action to be taken for the current crop in most cases. The N:S ratio is currently preferred by ADAS for grass at second cut silage stage to assess whether a field needs sulphur fertiliser. It has also been used to assess cereal grain at harvest.

Chapter 8

TRACE ELEMENTS

ALL CROPS need an adequate supply of the trace elements or micro-nutrients: iron, manganese, boron, copper, zinc, molybdenum and chlorine. Crop removal of these trace elements is very small compared to the major nutrients (Table 8.1) and most UK soils have adequate supplies without the need for specific fertiliser applications. Various fertiliser materials contain trace elements in appreciable quantity and help to maintain soil levels. Liming materials contain varying amounts of essential trace elements. Phosphate fertilisers contain these elements as impurities, particularly iron, copper and zinc. There is a recycling of trace elements back to the land in slurry and farmyard manure. Sewage sludges contain trace elements which can be utilised so long as over-application and toxicity are avoided. Some fungicides contain manganese, zinc or copper. Atmospheric inputs are also appreciable (see Chapter 2). Crops vary considerably in their suscepti-bility to trace element deficiencies. While individual deficiencies are common in particular crops on specific soil types, manganese deficiency is the only one that occurs widely. Crops rarely suffer from more than one deficiency at a time, but iron and manganese deficiencies

Table 8.1. Amounts of trace elements removed in crops

	% DM at harvest	Fe	Mn	Cu	B	Zn	Mo
				g/t fresh material			
Cereal–grain	85	40	25	4	0.8	25	0.3
–straw	85	40	60	2.5	6	15	0.3
Sugar beet–roots	22	20	7	1	3	4	0.1
–tops	16	30	8	1	6	3	0.1
Potato–tubers	22	4	4	2	0.7	4	0.05
Grass–silage	20	30	20	2	2	10	0.3
–hay	85	120	130	6	7	40	2
Kale	15	5	10	1	5	5	0.2

can occur together in fruit crops and copper and manganese deficiencies in cereals.

ADAS experiments have shown that a yield or quality benefit from the application of trace elements is unlikely unless a specific deficiency has been diagnosed. There is no economic justification for the routine application of trace element mixtures to field crops not showing deficiency symptoms. As deficiencies normally occur individually, specific treatments should be applied to the soil or crop as appropriate to remedy a problem in a particular field. This demands an accurate diagnosis of the cause of the crop symptoms. Often soil or plant analysis is worthwhile to support visual diagnosis and once a deficiency is diagnosed, recommendations for the prevention of the problem in future crops can be made.

Animals require an adequate supply of iodine, selenium and cobalt in their feed as well as the trace elements required by plants. The exception is boron which is essential for plants but not animals. The uptake of these trace elements by grass is important in maintaining an adequate supply for the grazing animal, particularly where no supplementary feeding is taking place.

IRON

Iron is taken up by plant roots as ferrous iron Fe^{2+} or as iron chelates. In most situations, reduction of ferric iron Fe^{3+} to ferrous iron Fe^{2+} takes place in the rhizosphere due to a lower pH in this environment. There is considerable variation between species in achieving an adequate supply of Fe^{2+} in the root rhizosphere. Several other cations have been shown to compete with Fe^{2+} uptake.

Iron is involved in several plant processes. These include a number of enzyme systems as well as a probable role in chlorophyll formation. Deficiency symptoms are generally quite specific. Most species show a general chlorosis of the younger leaves, with the veins remaining green (see Plate C.9). As symptoms intensify, young leaves may turn completely white followed ultimately by tip dieback in many fruit and ornamental species.

Soil Iron

All UK soils contain appreciable levels of total iron, up to 5 per cent or more. It is present as iron compounds and in clay minerals. The amount of total iron that is present in soluble form is extremely low over the pH range 6.5 to 8.0 at which most crops are grown.

Soluble iron only occurs in appreciable amounts at low pH or under the reducing conditions associated with waterlogging. A major source

Plate C1. Acidity in sugar beet Agricultural Lime Producers Council

Plate C2. Soil pH indicator colour range

Plate C3 (*top*). Nitrogen deficiency in winter barley

Plate C4. Phosphorus deficiency in forage maize

of available iron in soils are the various chelates associated with the soil organic matter. These are generally more available for plant uptake than the inorganic compounds.

Crop Response

Under UK conditions, iron deficiency in annual crops is rarely of practical significance. Symptoms may be seen in sugar beet and occasionally in other arable crops. Practical problems are limited to fruit and nursery stock. Deficiency only occurs on calcareous soils, particularly if poorly drained. Most fruit crops may be affected. The most susceptible are pear, raspberry and blackberry. Gooseberries and blackcurrants are fairly resistant. Numerous species of trees and shrubs are susceptible, particularly the Ericaceous ones. Not only do species vary in susceptibility, but variation also occurs between varieties of, for example, grapes and apples.

Most of the problems seen in the UK are lime induced chlorosis. It has been shown that iron deficiency on calcareous soils is generally due to bicarbonate ions, HCO_3^-. This ion reduces uptake and translocation of iron in the plant. Some studies show that iron uptake is not depressed by high bicarbonate concentration in the soil, but translocation to the young leaves is restricted. The iron is present in the plant but is chemically immobilised and so cannot carry out its normal function, resulting in symptoms of lime induced chlorosis.

High soil levels of bicarbonate are most likely when the soil carbon dioxide content is above normal. If soil structure and aeration are poor, carbon dioxide will accumulate in the soil reacting with calcium carbonate to produce bicarbonate ions. This fits the practical observation of more iron deficiency problems following springs with above average rainfall.

High levels of other heavy metal cations in the soil can induce iron deficiency in many annual as well as perennial crops. In particular, high levels of copper, nickel or zinc, either singly or in combination, can induce severe chlorosis. This is occasionally seen where excessive amounts of contaminated sewage sludge have been applied (see Chapter 11).

Diagnosis and Treatment

Visual symptoms are usually an adequate basis for diagnosis, particularly in broad leaved species. Leaf analysis for total iron is usually misleading as deficiency is due to reduced iron activity within the leaf. Some researchers have claimed success with various techniques for measuring soluble iron in plant material. Others have used enzyme activity measurements. None of these methods has been commonly adopted in the UK, probably due as much to the lack of need as to the inadequacy of the techniques.

When deficiency occurs, treatment is more difficult than for any other nutrient deficiency. Treatment is normally only attempted in severe situations. Foliar applications of inorganic ferrous salts may be useful, but have a high risk of causing leaf scorch. Foliar applications of the iron chelate EDTA (see page 110) are more successful but several applications are often needed to show a worthwhile control.

The most successful but most expensive treatment is the use of a soil applied chelate. This must be sufficiently stable in the soil so that the iron in the chelate is not rapidly replaced by calcium and the effectiveness lost. Several studies have shown that the chelating agent Fe–EDDHA is suitable for this purpose and it is commercially available for soil applications to fruit and nursery stock. Other chelates such as Fe–DTPA and the cheaper Fe–EDTA are much less successful.

Rates of Application

Soil applied iron is recommended at 60 g Fe–EDDHA per tree for orchard crops applied in February. Half this rate may be adequate in subsequent years. For close spaced crops apply 8 g per square metre to the rooting area annually.

For foliar treatment, apply Fe–EDTA at 1 kg/1000 l/ha plus wetter. Several applications at fourteen-day intervals may be needed. Foliar treatment may result in green spotting of leaves rather than overall leaf colour improvement. Leaf colour takes about five days to improve following application.

MANGANESE

Manganese is the trace element most commonly deficient in field crops in the UK and crop failure can occur if crops are not treated. The most severe problems occur on peaty and organic soils with a pH above 6.0. Deficiency is common but generally less severe on sands. On mineral soils, the deficiency is associated with a soil pH of 6.5 or above. It is frequently induced by overliming. Symptoms are often transient and may disappear following rain. Cereal crops following ploughed-out permanent grassland often show deficiency.

Plant Metabolism

Manganese is generally taken up as the divalent cation, Mn^{2+}. Mobility of manganese in the plant seems to vary with species. Most show deficiency symptoms in the older leaves, suggesting that translocation to the younger leaves has occurred. A few such as potatoes show deficiency in the younger leaves first. Deficiency is commonly diagnosed by visual symptoms, but these vary considerably between species and are summarised for the commonly affected crops in Table 8.2. Manganese is involved in many plant functions, especially in enzyme systems. It is involved in photosynthesis and change of leaf colour is often the first visual symptom of deficiency.

Soil Manganese

Like iron, the total manganese content of UK soils is generally adequate. Deficiency occurs because insufficient is chemically available for plant growth. Manganese occurs in several primary minerals as well as on clay mineral surfaces. Divalent Mn^{2+} occurs in soil solution, on clay mineral surfaces and in organic matter complexes. Manganese also occurs in the soil in several other oxidation states, principally in tri- and tetravalent forms. The equilibrium between these various states depends on oxidation–reduction changes in the soil.

A fraction called easily-reducible manganese is often measured by soil analysis and largely determines the manganese available for plants. Unfortunately this level fluctuates as soil temperature and water content change, influencing microbial activity and oxidation conditions. High levels of soluble Mn^{2+} occur in acid soils. Waterlogging often increases the amount of manganese in soluble, reduced Mn^{2+} form. In very sandy soils low in organic matter, absolute deficiency of manganese can occur. Generally, deficiency occurs on high pH soils which have insufficient clay to maintain an adequate level of exchangeable Mn^{2+}.

Diagnosis and Treatment

Most crops are susceptible to manganese deficiency, although symptoms are uncommon in grasses, rye, field beans and gooseberries. Diagnosis is often possible by visual symptoms (see Table 8.2) and leaf analysis is a reliable guide when in doubt. A leaf level below 20 mg/kg [dry matter (DM) basis] indicates deficiency for a wide range of species, correlating well with visual symptoms and crop response (see Figure 8.1). ADAS and Brooms Barn work has shown that yield response does not generally occur in the absence of visual symptoms.

While soil analysis for exchangeable or easily-reducible manganese is used by some commercial companies in the UK it is not used for advisory purposes by ADAS because:

1. Soil levels fluctuate with change in soil oxidation state as described earlier.
2. Foliar diagnosis is an adequate and early enough basis for taking decisions on foliar spraying.

Table 8.2. Symptoms of manganese deficiency

Cereals	Cereals do not show leaf symptoms before the three-leaf stage. Field patches show as areas of yellow or pale green, floppy growth. Characteristic symptoms are first seen on the middle third of the leaf. The tip remains green, even if the leaf breaks in the middle and hangs down.
Oats	Very susceptible, showing interveinal yellowing with grey/buff streaks (grey speck) which turn necrotic in severe cases.
Wheat	Rows of interveinal white streaks occur (see Plate C11).
Barley	Spring barley shows rows of interveinal brown spots (see Plate C12). Barley under stress may show an overall leaf spotting not associated with manganese deficiency. Winter barley symptoms resemble those on wheat.
Sugar beet	Very susceptible, beet may show symptoms at the two-leaf stage. Older leaves develop an overall interveinal yellow mottle and the leaf margins curl inwards giving a triangular shaped, erect leaf habit (see Plate C13).
Potatoes	This crop shows initial paleness of younger leaves, followed by blackish-brown spots along the veins in some varieties. These are best seen on the underside of the leaf. Symptoms may be confused with overall stress induced spotting in some varieties.
Oilseed rape	Older leaves show interveinal yellowing (see Plate C14) particularly near the leaf margins. The symptoms are easily confused with magnesium deficiency although this usually shows some reddening.
Peas	Occasionally seen in the older leaves as interveinal yellowing, dried peas show marsh spot, an internal brown discolouration in the centre of the pea which is not visible until the pea is split.
Redbeet	Loss of green colour produces leaves with an overall dull red colour. The leaf margins curl inwards. Field patches have a very distinctive overall dull red colour.
Dwarf beans	Very susceptible, the older leaves show a strong interveinal yellow mottle while the veins remain green.
Carrots	Deficiency occurs as light green patches in the field, which are difficult to distinguish from other nutrient deficiencies.
Onions	The main symptom is a pale green leaf. Older leaves show yellow mottling in severe cases.
Celery	Older leaves develop a strong interveinal yellowing especially near the leaf margins.
Fruit	Most species show interveinal yellowing of leaves which begins at the leaf margins and gradually extends towards the midribs leaving a band of green along the vein.

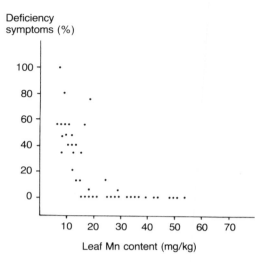

Figure 8.1. Manganese deficiency symptoms and leaf manganese
content of sugar beet
Source: Brooms Barn Experimental Station

As manganese deficiency is generally due to low availability in the soil, soil treatment is not economically worth while. Some podzol soils have responded to soil applications, where the total soil manganese level is inherently low. In most situations chelated applications would be needed, which are not as cheap or as effective as foliar sprays.

Response to a foliar spray of manganese is generally rapid and reliable. The spray should be applied as soon as symptoms are diagnosed. A preventive spray should be applied on fields with a history of manganese deficiency, as soon as crop cover is adequate. Autumn spraying is worth while on early-sown winter cereals on very susceptible soils. This will reduce the risk of frost damage, commonly seen on deficient crops in severe winters.

The standard recommendation is 9 kg/ha of manganese sulphate plus wetter in at least 250 l of water. Some crops may need more than one spray. Foliar scorch may occur if crops under water stress are sprayed in bright sunshine. Fruit crops should receive 4 kg/ha in 1,000 l of water. An insurance application of 5 kg/ha should be applied to dried peas at the flowering stage if grown on sandy or organic soils with a pH above 6.5 or 6.0 respectively.

Recent ADAS experiments have compared different rates of manganese sulphate and chelated manganese on a number of sites. While chelated manganese is generally more effective when compared on a weight of manganese basis, the lower manganese content and higher cost of chelated forms generally mean that manganese sulphate is the preferred choice for most circumstances. These experiments also show that 5 kg/ha of manganese sulphate is an adequate rate for most circumstances.

Sugar beet often shows deficiency symptoms before there is sufficient leaf cover to take up foliar applied manganese. Sugar beet seed with manganese oxide incorporated in the seed pellet is available in the UK and should be used on fields with a history of this deficiency. An additional foliar spray may be necessary in June, but by this time leaf cover should be adequate for this to be effective. Manganese is a heavy metal, so 1 kg of manganese pelleted seed contains only about half the number of normal, clay pelleted seeds.

BORON

Boron deficiency is far less common than manganese deficiency. It occurs mainly on sands and light loams. Soil boron levels are also low in soils derived from the shales of Wales and south west England. Soil analysis provides a useful guide to the likelihood of boron deficiency. Deficiency is more likely on soils over pH 6.5 especially if they have been recently limed. Since boron is water soluble and readily leached from sandy soils, deficiency is more common following a wet winter and spring. Summer drought also increases the likelihood of deficiency.

Plant Metabolism

Boron remains the least understood of the nutrients as far as detailed explanation of its uptake and function is concerned. Uptake is generally considered to be as undissociated boric acid, H_3BO_3. It is relatively immobile in plants, phloem transport of boron being poor, so distribution depends on the rate of transpiration.

Deficiency shows first in the growing point, usually when the crop transpiration rate is limited by water stress. Boron is involved in the development of root and shoot growing points and any deficiency results in disruption of meristematic growth. Boron has been shown to be involved in a number of specific biochemical functions; however, many of these observations need further verification.

Detailed symptoms of deficiency for the major susceptible UK crop plants are given in Table 8.3. Most show damage to the growing point often with petiole cracking where transpiration has been interrupted. Root crops show various symptoms in the storage organ that are related to boron supply.

Table 8.3. Symptoms of boron deficiency

Sugar beet	Leaves and petioles are brittle. The upper surface of the midrib often shows transverse cracking. Young leaves develop an uneven yellow mottle and may die (see Plate C10). Secondary growth gives a rosette appearance to the crown and internal breakdown of the tissue beneath the crown occurs. Similar hollow crown symptoms may occur which are not due to boron deficiency.
Swedes and turnips	These show internal, brown, water-soaked areas in the root, which are only visible after cutting. The leaves and growing point are rarely affected.
Oilseed rape	The crop very rarely shows the characteristic symptom of loss of growing point in the UK. Flowering and podset may be affected but symptoms are not easily diagnosed.
Lucerne	Lucerne shows yellowing and reddening of the upper leaves. The upper stem is often stunted with numerous short side branches.
Redbeet	Internal hard black necrotic areas develop in the root which are only visible after cutting. Symptoms are often worse after cooking. Leaf symptoms are rarely seen.
Carrots	Carrots exhibit a symptom known as '5 o'clock shadow' which shows as a dark colouration on the surface of the central portion of the root either as diffuse areas or tiny spots.
Cauliflower and cabbage	Plants show initial leaf curling. Leaves become brittle and mottles occur around the margins. Small blister-like swellings appear on the stem and lower surface of the leaf stalk. Cauliflower may show browning of the curd. The stem is often hollow and discoloured internally near the top. However most hollow stem and stem cracking seen in brassicas is physiological and is not due to boron deficiency.
Celery	Brown cracks develop along the ridges of the outer surface of the petioles and the tissue turns brown. The cracked surface may break and curl. Severe deficiency shows as a blackening of the heart which may be confused with calcium deficiency.

Soil Boron

Various soil minerals contain boron but most is in forms unavailable to plants. Available boron occurs mainly as the undissociated boric acid, which does not ionise at neutral soil pH levels. At high soil pH, borate $B(OH)_4^-$ occurs which is strongly adsorbed by sesquioxides and some clay minerals. Some boron is also held in organic combination, particularly under slightly acid conditions. Boron deficiency only occurs on low organic matter soils. It is generally only seen at high soil pH levels because the borate formed under alkaline conditions is strongly adsorbed and very little is available to plants.

Boron in soils as boric acid is readily leached. For this reason it is not possible to build up soil levels of available boron using normal soluble boron fertilisers. During the winter, excess rainfall will remove soluble boron remaining in the soil. Soils containing moderate amounts of either clay or organic matter usually provide adequate boron for crop growth.

Diagnosis and Treatment
Boron deficiency only occurs in the UK when one of a small number of susceptible crops is grown on a potentially deficient soil. Deficiency is most common in sugar beet, swedes and carrots. Cereals, grasses, potatoes, peas and beans have not been recorded as boron deficient in the UK. General symptoms are death of the growing point which then stimulates growth of lateral buds. Petioles become necrotic and leaves chlorotic and brittle.

Soil and plant analysis can be used to complement visual diagnosis, but once symptoms of deficiency are seen it may be too late to apply effective remedial treatment. Soil analysis should be used to assess the boron need for susceptible crops grown on sandy soils. Where the hot water soluble method for available boron used by ADAS indicates levels of 0.5 mg/l or less, boron should be applied before growing all susceptible crops. Between 0.5 and 1.0 mg/l deficiency is possible, but it is unlikely above 1.0 mg/l. Between 0.5 and 1.0 mg/l boron should be applied to carrots as an insurance dressing if the pH is above 7.0.

Leaf analysis can be used to confirm visual symptoms, but the poor mobility of boron in the plant makes sampling difficult. While deficiency is often assumed if the leaf level is below 20 mg/kg (DM basis), a figure of ten or below is generally needed in the youngest leaves for definite diagnosis. Soil analysis is commonly of more diagnostic use. Diagnosis in root crops can be confirmed by analysis of the root peel: levels below 20 mg/kg are considered deficient.

Soil applications of boron should be applied before sowing and thoroughly cultivated into the ground. Borax (20 kg/ha) may be used but it is difficult to apply evenly because of the small quantity involved. It is only sparingly soluble and cannot be applied in solution. An alternative method is to use a more soluble material such as Solubor. This should be applied at 10 kg/ha in an appropriate quantity of water, sprayed directly onto the soil and thoroughly mixed in. The most straightforward technique is to apply a boronated fertiliser.

Susceptible fields generally need boron for each susceptible crop in the rotation, but not more frequently than every third year. If circumstances only allow time for a foliar spray, this must be applied as early as leaf cover allows. Apply 5 kg/ha of Solubor in at least 250 l of water. Soil or foliar boron should not be applied to unresponsive crops, particularly cereals, potatoes or french beans. Boron toxicity symptoms are likely if these crops are treated.

COPPER

Copper deficiency has only been diagnosed in a few specific soil situations in the UK. It occurs on organic soils mainly in the Fens and on leached sandy soils, particularly reclaimed heathland. The other main area of recorded deficiency is on the shallow, organic chalk soils of southern England.

Plant Metabolism
Copper is taken up as the cation Cu^{2+}. It is not very mobile within the plant, particularly in deficient crops. In high copper soils, copper is concentrated in the plant roots and little is translocated to the shoot. Copper performs a number of essential biochemical functions. It is involved in several enzyme systems and also in photosynthesis.

Soil Copper
Copper occurs predominantly as the divalent cation Cu^{2+}, either as an exchangeable cation on clay surfaces or complexed with organic matter. Copper is held strongly by the soil organic matter which plays a major part in determining whether adequate copper is available for crop growth. Exchangeable copper on the clay minerals is held very tightly. Negligible leaching of copper occurs from all except very sandy soils.

Diagnosis and Treatment
Deficiency is generally restricted to wheat and barley, though cases of deficiency in sugar beet, onions and pears have been recorded in this country. Symptoms in cereals are not usually seen until the end of tillering and often not before flag leaf emergence. Yellowing of the tips of the youngest leaf is often followed by spiralling of the leaves. Symptoms may easily be confused with drought stress, herbicide or frost damage. Ears are sometimes trapped in the leaf sheath and those that emerge have white tips that do not fill with grain. Awns of barley become white and brittle (see Plate C.15). Blackening of the ears and straw occurs in copper-deficient wheat on organic chalk soils. This symptom has not been recorded in cereals on sands or peats. Recent work has shown that empty florets in cereals may result from copper

deficiency causing failure of the anthers to develop and pollen sterility. This is usually seen in florets at the terminal end of the ear.

Soil analysis is often useful to confirm visual diagnosis of copper deficiency, especially where symptoms are confused by those of drought. Visual diagnosis is commonly complicated by manganese deficiency symptoms on peats and sands. Plant analysis is generally of little help in diagnosis except in fruit. In cereals, analysis of grain at harvest has been shown to give a good indication of deficiency, although retrospective. Copper levels below 2 mg/kg (DM basis) in the grain are generally considered deficient. Soil analysis enables preventative treatment to be carried out where the soil available copper level, using the ADAS method of EDTA extraction is below 1.0 mg/l. Deficiency is possible on peaty and organic soils between 1.0 and 2.5 mg/l. The higher the soil organic matter content, the higher the threshold soil copper level. Visual response to a copper spray usually occurs in seven to ten days and this is the best confirmation of deficiency. Deficiency has been shown more frequently following a brassica, especially kale, on the organic chalk soils of southern England.

A foliar spray of 2 kg/ha copper oxychloride or cuprous oxide fungicide in 500 litres of water is recommended for cereals. Alternatively a chelated copper product may be used with less risk of crop scorch. Fields that have a history of copper deficiency should be sprayed in the spring as soon as full crop cover is established. Do not wait for symptoms to appear in crops on known susceptible fields. For these situations, a soil application of 30–60 kg/ha of copper sulphate crystals can be applied. This will prevent deficiency for a number of years. For pears, apply 2 kg/ha of copper fungicide in 1,000 litres of water in late May.

ZINC

Deficiency has only been recorded a few times in the UK. The crops affected have been apples and forest nursery stock. It occurs on sandy soil with high pH and phosphate levels.

Plant Metabolism

Zinc is taken up as the divalent cation, Zn^{2+}. Movement within the plant is poor, especially under deficient conditions. Deficiency symptoms occur in the growing points of most species. It is commonly observed overseas that zinc deficiency symptoms are associated with a high soil phosphorus level. In some cases high phosphorus within the plant has been shown to induce zinc deficiency symptoms. The main function of zinc is in plant enzyme systems.

Soil Zinc

Zinc, mainly as the cation Zn^{2+}, occurs both on the exchange surfaces and complexed with organic matter. Low levels of zinc in the soil solution are usually associated with high pH, calcareous soils. Organically bound zinc is a major source of available zinc for crop uptake.

Diagnosis and Treatment

Zinc deficiency generally shows an interveinal chlorosis of the leaf. Apples show restricted spur growth, with a characteristic rosette of small leaves at the tip. In maize, chlorotic bands form on either side of the midrib of the leaf. Both leaf and soil analysis are used in other countries for prediction of the need for zinc application and in the diagnosis of deficiency. In the UK leaf levels below 15 mg/kg Zn are generally considered deficient. As deficiency is so rare in the UK, little work has been done on soil analysis methods for deficiency prediction.

Treatment of zinc deficiency in the UK is limited to experience of foliar applications. Apples have been successfully treated with a winter spray of zinc sulphate at 4 per cent weight/volume (w/v) or at 0.1 per cent w/v spray at green cluster or petal fall.

MOLYBDENUM

Molybdenum deficiency is associated with acid soil conditions and is not generally a problem on adequately limed fields. Molybdenum is the only essential plant trace element that is less available at low pH.

Plant Metabolism

Molybdenum is taken up as the molybdate anion, MoO_4^{2-}. It is only moderately mobile in the plant. Molybdenum is required in much smaller amounts than the other trace elements. The known functions of molybdenum in plants are limited to two enzymes, nitrogenase and nitrate reductase. Nitrogenase also occurs in free living and symbiotic nitrogen fixing micro-organisms. Molybdenum deficiency leads to nitrate accumulation in plants as the enzyme activity to convert the nitrate to nitrite is restricted. Symptoms of deficiency resemble nitrogen deficiency at least in the initial stages because nitrogen is accumulated but not metabolised. In legumes, molybdenum deficiency can give very similar symptoms to nitrogen deficiency, as both leaf nitrate reduction and nitrogen fixation may be inhibited.

Soil Molybdenum
Soil parent materials vary considerably in their molybdenum content. No soils in the UK are recognised as being generally deficient in available molybdenum at pH levels normal for crop growth. Molybdate is adsorbed by sesquioxides and clay minerals rather like phosphate and is most strongly held at low pH levels. Raising soil pH by liming will often control deficiency by reducing the strength of adsorption. Some molybdenum is held by the soil organic matter.

Diagnosis and Treatment
Cauliflower and lettuce are the only field crops commonly affected. Deficiency is far more common during the plant raising stage in peat compost than after planting out in the field. Both crops show characteristic chlorosis of the leaf edges. In severe field situations, cauliflower produces the characteristic 'whiptail' symptoms where growth of the leaf lamina is severely restricted giving narrow strap-like leaves (see Plate C.16). The remaining lamina is irregular in outline and puckered. Young plants show yellowing and inward cupping of the leaves often followed by blindness. Molybdenum deficiency has not been recorded in legumes in this country. Lucerne and clover suffer deficiency in other countries.

Soil analysis levels below 0.1 mg/l available molybdenum measured by the ADAS Tamm's oxalate method are generally considered deficient for sensitive crops. Leaf levels below 0.1 mg/kg are considered deficient. Accurate analysis at these low levels requires considerable care. Leaf analysis is generally the more reliable for confirming deficiency.

Where needed, soil treatment with sodium or ammonium molybdate at 0.03 g per square metre is recommended or a foliar drench at 0.25 g/l if deficiency occurs in the plant raising stage. Routine foliar application is often worthwhile for peat compost raised cauliflower and lettuce before planting out. Liming to pH 6.5 will generally prevent field deficiency.

CHLORINE

Chlorine is an essential element required in small quantities. It is widely distributed and field deficiency has not been recorded in the UK. Chlorine occurs in soils as the chloride anion, Cl^-. It is not adsorbed so leaching readily occurs. The chloride contribution from the atmosphere is adequate to meet crop needs even without the considerable chloride applied in many fertilisers.

IODINE

Iodine is an essential trace element for animals. It probably occurs in soils as both the anions, iodide I^{2-} and iodate $(IO_3)^{2-}$. There is little work relating grass iodine content to soil characteristics. Liming is likely to reduce crop content, particularly above pH 7.0. In practice most pastures contain too little for breeding and lactating animals and supplementary iodine should be fed. Applications to soil or crop are less satisfactory.

SELENIUM

Selenium is an essential element for animal nutrition. Selenate, $SeO_4{}^{2-}$, is the most important form that is taken up by plants. Both toxicity and deficiency in grazing animals have been found, depending on the selenium levels in the soils. In the UK, Carboniferous and Ordovician slates, shales and limestones are most likely to give toxicity. Deficiency is most likely on granite and sandstone soils. Selenium is more available for plant uptake at high pH. Application to soil or crop is not recommended due to the risk of toxicity problems.

COBALT

Cobalt is required by animals and by nitrogen fixing micro-organisms. The requirement by the latter is so small that deficiency in the UK has not been recorded. Cobalt deficiency in animals, particularly sheep, is quite common on low cobalt soils in the UK. Cobalt occurs in available form in soils as the divalent cation, Co^{2+}. For animals, herbage needs to contain 0.08 mg/kg or more of cobalt. Deficiency is likely on highly leached, sandy soils, soils derived from acid igneous rocks or in highly calcareous or peaty soils at high pH levels. Values below 0.1 mg/l in the soil by the ADAS method for available cobalt are considered deficient.

Cobalt is the one animal trace element which is appropriate for treatment by application to pasture. In deficient situations apply 2 kg/ha of cobalt sulphate to grassland in the spring and this will provide satisfactory herbage levels for at least one season. As the increase in grass cobalt level from this treatment is considerable, only part of a field need be treated each year.

Chapter 9

FERTILISER MATERIALS

ALTHOUGH A few chemical compounds dominate the UK inorganic fertiliser market, there is a wide range of fertiliser materials sold. The effectiveness of these materials for crop production is discussed in this chapter. The various commercial products and regulations governing them are discussed in Chapter 10. The majority of this chapter is devoted to the nitrogen and phosphorus sources available in the UK. This is followed by consideration of the more limited range of forms in which the other major nutrients and trace elements are available. Choice of material will generally be governed by economics. If a particular material is to be used, it must be both agronomically suitable and economically competitive.

NITROGEN FERTILISERS

As described in Chapter 3, most nitrogen is taken up by plants as nitrate. All inorganic nitrogen compounds applied to the soil are converted to nitrate by the soil micro-organisms. The main inorganic forms of nitrogen fertilisers are ammonium (NH_4^+), nitrate (NO_3^-) and ureic (NH_2). Commonly available materials are listed in Table 9.1 and discussed below.

Ammonium nitrate (33.5–34.5 per cent N)
Ammonium nitrate is the dominant, straight nitrogen fertiliser used in the UK. Plate 9.1 shows ammonium nitrate prilling towers. Half the nitrogen content is ammonium–N; the other half is nitrate–N. It can be applied to the soil surface under a wide range of conditions with minimal direct losses of gaseous nitrogen by volatilisation of ammonia to the atmosphere.

In Great Britain, stored ammonium nitrate is no longer considered an unacceptable fire hazard. It will intensify a fire involved with other

Table 9.1. Main nitrogen fertilisers

	Approximate nitrogen content % N
Ammonium nitrate, NH_4NO_3	34.5
Ammonium nitrate–lime mixture, $NH_4NO_3 + CaCO_3$	26
Anhydrous ammonia, NH_3	82
Aqueous ammonia, $NH_4OH + NH_3$	26
Urea, $CO(NH_2)_2$	46
Ammonium sulphate, $(NH_4)_2SO_4$	21
Calcium nitrate, $Ca(NO_3)_2.NH_4NO_3$	15.5
Ammonium nitrate sulphate, $NH_4NO_3.(NH_4)_2SO_4$	26
Calcium cyanamide, $CaCN_2$	21
Nitrate of soda, $NaNO_3$	16(26Na)
Potash nitrate, $NaNO_3 + KNO_3$	15(15K_2O, 18Na)

combustible materials, and under no account should straight ammonium nitrate be stored near straw or other highly inflammable organic material.

Ammonium nitrate–lime mixtures (25–26 per cent N)
Up to the 1960s all ammonium nitrate was sold mixed with calcium carbonate. This was before the advent of prilled ammonium nitrate which produced a physical form suitable for storage and spreading. The mixtures are still available in the UK but generally cost more per kg of nitrogen than straight ammonium nitrate. They contain calcium carbonate, which partially but not completely compensates for the acidifying effect of the ammonium nitrate. From a crop use viewpoint, straight ammonium nitrate and ammonium nitrate–lime mixtures may be considered identical. This mixture is sometimes known as calcium ammonium nitrate (CAN).

Ammonia
Ammonia is produced for agriculture as either anhydrous ammonia (82 per cent N) or aqueous ammonia (21–29 per cent N). As the name implies, anhydrous ammonia is a liquified gas which must be stored in containers pressurised up to 20 bars until released through injector tines into the soil. Aqueous ammonia is anhydrous ammonia dissolved in water under slight pressure. It is therefore more bulky, but needs less expensive and sophisticated storage and application equipment. Together both forms currently constitute only a small proportion of the UK nitrogen market.

When ammonia is injected into the soil, the ammonium ions are adsorbed by the clay and organic matter. Subsequently the ammonium

Plate 9.1. Ammonium
nitrate prilling towers
ICI plc

is nitrified by soil micro-organisms to nitrate. High concentrations of
anhydrous ammonia injected in a narrow band will partially sterilise
the soil and delay the conversion process until biological activity is
re-established. Whether injected for arable crops or grassland, the soil
conditions must be satisfactory or loss of ammonia gas can be considerable. When walking behind an injection machine there should be
no smell of ammonia. Depth of injection will generally need to be
15 cm for anhydrous and 10 cm for aqueous ammonia. Injection
during seedbed cultivation can generally be achieved satisfactorily on
any soil type. The main problem is the wheelings resulting from the
application. Efficiency of injection on grassland or as a topdressing
may be reduced on heavy soils when either too wet or too dry, due to
poor slit sealing behind the machine. Ammonia loss due to poor
retention by the soil may occur on very sandy or stony soils, particularly if low in organic matter or under dry conditions.

In general under UK conditions, ammonia needs to be cheaper per kg of nitrogen than ammonium nitrate before it is attractive as an alternative. For most situations it offers no advantages and provides more problems of storage and application. Aqueous ammonia is considerably easier to manage than anhydrous ammonia. When used under satisfactory conditions of soil and timing, ammonia is as effective a nitrogen source as ammonium nitrate.

When using either source of ammonia strict operator safety precautions must be followed at all times. For anhydrous ammonia the code of practice prepared under the guidance of the Chemical Industries Association should be followed. Aqueous ammonia is sometimes stored in flexible butyl containers which puncture relatively easily. These should be sited away from watercourses and vandals.

Urea (45–46 per cent N)
Like ammonia, urea is generally only attractive to UK farmers if available at a lower price per kg nitrogen than ammonium nitrate. Urea is sold in prilled form or in non-pressure solutions. In the soil, urea $CO(NH_2)_2$ is converted by the enzyme urease to ammonium carbonate. The main agronomic limitation of urea results from the instability of this intermediate breakdown product. When topdressed under some conditions it can release free ammonia gas which is lost to the atmosphere. When incorporated into the seedbed, there is a risk of root damage to seedlings.

As long as precautions are taken to minimise these problems urea can often be as effective a nitrogen fertiliser as ammonium nitrate. Straight urea should not be combine drilled. Urea-containing compounds should only be combine drilled at a modest rate of nitrogen, especially in dry seedbeds and on sandy soils. When urea is broadcast and incorporated, problems are minimised.

Soil type and weather following application are the two main factors determining the loss of ammonia after topdressing. The risk of appreciable ammonia loss is most likely on very sandy or light, chalky soils. On these soils, perhaps 10 per cent or more of a urea nitrogen topdressing will be lost during some years. The risk is of little practical significance on medium and heavy soils. Loss is most likely to occur when urea is topdressed onto moist soil and the following few days are warm and dry; these conditions are unlikely before April. If rain follows within a day of application, losses will be minimal. There is some evidence that the effectiveness of a particular application of urea will decrease as the level of nitrogen applied increases—particularly above 100–120 kg/ha N.

Urea is as good as ammonium nitrate for mid February–early March application to winter wheat. The main April application is more at risk and choice will depend on soil type as discussed earlier. To reduce the risk of ammonia loss the main application should be split, perhaps half in early April and half two weeks later at around early stem extension. For winter barley the same principles apply as for winter wheat. As topdressing will generally be completed earlier, the risk of loss is less. Urea applied late February–early March to winter oilseed rape is likely to be as good as ammonium nitrate. Recent work comparing ammonium nitrate and urea for silage showed variable results for first cut. Urea was generally less satisfactory unless rain followed application, but the differences were usually small. As anticipated, urea was less satisfactory when applied in May for second cut. Yield penalties of up to 25 per cent were shown on some sites. There is a case for splitting the application of urea for first cut silage, where its use is considered appropriate.

When urea is being applied for any crop in conditions where there is a risk of loss, the use of 10 per cent extra nitrogen will minimise the risk of a yield penalty in most conditions. On sandy and chalk soils, the yield loss from urea cannot always be made good by increasing the application. Because urea is slower acting than ammonium nitrate, it should be applied a few days earlier than ammonium nitrate would be.

Prilled urea is both less dense and of smaller prill size than ammonium nitrate. In spinning disc and oscillating spout machines, bout widths must be narrower and spreading will be more affected by wind. Pneumatic machines will need recalibration but otherwise present fewer difficulties.

Foliar sprays of urea are less likely to cause crop scorch than a similar concentration of ammonium nitrate. Urea may contain biuret formed during manufacture. Biuret is toxic to plants if present in high concentration. As long as the urea has less biuret than the statutory maximum of 1.2 per cent, it will cause no toxicity problems.

Sulphate of ammonia (21 per cent N)
Ammonium sulphate is sometimes available in the UK. It is an effective nitrogen source, but causes greater soil acidification per kg of nitrogen applied than other materials. As with urea, topdressing may result in some ammonia loss.

Calcium nitrate (15.5 per cent N)
Fertiliser calcium nitrate is formulated with 5 per cent ammonium nitrate. It is not widely used in the UK, but it is useful if a mainly nitrate or a non-acidifying nitrogen source is required.

Ammonium nitrate sulphate (26 per cent N)
A double salt compound which offers no specific agronomic advantages compared to ammonium nitrate, unless sulphur is deficient. It has a high acidifying capacity.

Ammonium phosphates
Many compound fertilisers contain part or all of their nitrogen as mono or diammonium phosphate. If a compound contains a greater percentage of its nitrogen as ammonium than as nitrate, the extra nitrogen will usually be as an ammonium phosphate. Ammonium phosphates have no specific agronomic limitations. Diammonium phosphate (16:46:0) is sometimes available in the UK.

Calcium cyanamide (19–21 per cent N)
Calcium cyanamide ($CaCN_2$) breaks down in the soil to urea and subsequently to nitrate. The first stage of breakdown is relatively slow. This may be an advantage or disadvantage depending on the use. Some of the intermediate products are toxic to plants, hence its herbicidal and fungicidal properties. There is little experimental work on its use in the UK.

Nitrate of soda (16 per cent N, 26 per cent Na)
Nitrate of soda is a natural salt imported from Chile. Agronomically it is a satisfactory source of sodium and nitrate but is usually more expensive in the UK than other sources.

Chilean potash nitrate (15 per cent N, 10–15 per cent K_2O, 9–18 per cent Na)
One of a number of mixtures of naturally occurring salts, it is a relatively expensive nutrient source in the UK.

PHOSPHATE FERTILISERS

Phosphate fertilisers may be separated into those containing water-soluble phosphorus and those without part of their phosphorus content in water-soluble form. The majority of the arable and horticultural market needs water-soluble products; the main scope for less soluble products is on grassland. The materials sold in the UK are listed in Table 9.2 and discussed below. It should not be assumed that a water-soluble source will be more expensive per kg P_2O_5 than a less soluble product.

Table 9.2. Main phosphate fertilisers

	Approximate phosphate content *% P_2O_5*
Superphosphate	19
Triplesuperphosphate	47
Monoammonium phosphate	62 (12% N)
Diammonium phosphate	54 (21% N)
Ammonium polyphosphate	37 (11% N)
Dicalcium phosphate	40
Basic slag	5–22
Ground mineral phosphate	25–40
Aluminium calcium phosphate	32–34
Calcined phosphate	27–30

Superphosphate, (18–21 per cent P_2O_5)
Single superphosphate, normally sold in granular form as a straight, is by statute at least 93 per cent water-soluble P_2O_5. It is suitable for all crops on all soils. The main components are monocalcium phosphate and gypsum.

Triplesuperphosphate (47 per cent P_2O_5)
Triplesupers is a more concentrated form of superphosphate containing predominantly monocalcium phosphate. It is agronomically similar to single supers.

Ammonium Phosphates
Much of the phosphate in compounds is present as monoammonium or diammonium phosphates. These are very water-soluble materials. The phosphate is agronomically similar to the superphosphates. Diammonium phosphate (DAP) is sometimes available in this country (21:54:0). Monoammonium phosphate (MAP) (12:62:0) is not normally available.

Ammonium Polyphosphate (11:37:0)
The main commercial advantage of ammonium polyphosphate is that it allows higher concentration liquid fertilisers to be produced than those based on ammonium phosphates. Its use is almost entirely in liquid fertilisers. Agronomically it is the same as the ammonium phosphates.

Dicalcium Phosphate (40 per cent P_2O_5)
Although not soluble in water, dicalcium phosphate–P is almost all soluble in alkaline ammonium citrate. In finely divided form it acts

almost as quickly as water soluble sources. Many compounds produced in Europe, but not in the UK, contain a proportion of their phosphorus as dicalcium phosphate. As these are formulated as granules, this source of phosphate is likely to be less satisfactory in deficient situations or for responsive crops such as potatoes. For other maintenance situations it will be as good as water-soluble materials.

Basic Slag (5–22 per cent P_2O_5)
Basic slag, the mainstay of UK grassland fertilising for so long, is no longer available except for small quantities of low-grade, imported material. Apart from being suitable for use on arable crops and grassland, it has a neutralising value about two-thirds that of ground chalk or limestone on a weight basis. The loss of basic slag has been the commercial stimulus for the wide range of other insoluble phosphates that have been marketed in recent years.

Ground Mineral Phosphate (25–40 per cent P_2O_5)
Ground mineral phosphate is finely ground rock phosphate which has received no other treatment. Only the soft rock phosphates as distinguished by solubility in formic acid are satisfactory fertilisers. The best known soft rock is from Gafsa in Tunisia. The granular products now sold disintegrate on contact with water to release the powder.

Soft rock phosphate is best suited for use on grassland in the absence of natural or more particularly recently applied calcium carbonate. It is not as quick acting as water-soluble materials, but is well suited to maintenance applications on established grass especially in higher rainfall upland areas. Rock phosphates are not suitable for arable cropping where a soil pH of 6.5 or above is maintained.

Aluminium Calcium Phosphate (32–34 per cent P_2O_5)
Aluminium calcium phosphate is produced by calcining and grinding phosphate rock. ADAS experiments have shown it to be poor in the first year but it improves in the second and third years after application to grassland. It is suited to maintenance but not deficiency situations on grassland. Unlike rock phosphate, its efficiency is maintained in the presence of calcium carbonate.

Calcined Phosphate (27–30 per cent P_2O_5)
This product also called Rhenanian phosphate is produced by heating rock phosphate with soda ash and silica. It has given rather variable results in ADAS experiments and is usually more expensive than other insoluble products.

Mixtures
In recent years a range of mixtures of two different forms of phosphate have been sold. Phosphated slags containing basic slag and ground rock phosphate are no longer available due to lack of slag supplies. Currently several mixtures of ground rock phosphate and super-phosphate are being sold. These are suitable for establishment and maintenance on grassland as long as the conditions pertaining to rock phosphate use are upheld, and assuming the rock component is soft rather than hard. Partially solubilised rock phosphates are similar to these mixtures in agronomic effectiveness. These are formed by partial reaction of rock phosphate with acid resulting in a proportion of the phosphate in water-soluble form. Again their effectiveness depends on the type of rock phosphate used.

Chemical Analysis of Phosphate Fertilisers
In addition to total P_2O_5 content, various methods of chemical extraction are used to assess phosphate fertiliser materials (see Table 9.3). These methods are used to standardise the quality of materials of the same type. Field testing is needed to draw conclusions on the comparative agronomic qualities of differing types of phosphate fertilisers. Unfortunately variation in crop response between two materials of similar type is not always reflected in a difference in their extractable phosphorus content. Where non-water-soluble materials are used, details of crop response to the particular product should be studied to ensure that it is effective.

POTASH AND SODIUM FERTILISERS

All sources of potash and sodium commonly used in agriculture are completely water soluble, so choice depends mainly on cost. In some specific cases, choice may also be dependent on the other components in the material. The main materials are listed in Table 9.4.

Muriate of Potash (60 per cent K_2O)
Potassium chloride is by far the most important source of potash used in the UK. It is available as a straight, in granules or flakes and is used in both solid and liquid compounds.

Sulphate of Potash (50 per cent K_2O)
This is manufactured rather than mined so it is more expensive than the chloride. It is used in compounds for crops which are adversely affected by too much chloride.

Table 9.3. Chemical extractants used for assessing phosphate fertilisers

Extractant	Fertiliser
Water	superphosphates partially solubilised rock phosphate phosphate-containing compounds*
Neutral ammonium citrate	superphosphate phosphate-containing compounds*
Alkaline ammonium citrate	aluminium calcium phosphate calcined phosphate dicalcium phosphate
2 per cent citric acid	basic slags
2 per cent formic acid	soft ground rock phosphate phosphated slag rock phosphate
Total (mineral acid)	aluminium calcium phosphate basic slags partially solubilised rock phosphate soft ground rock phosphate phosphated slag rock phosphate

* Compounds which do not contain basic slags, calcined phosphate, aluminium calcium phosphate, soft ground rock phosphate or partially solubilised rock phosphate.
Source: The Fertiliser Regulations, 1977.

Table 9.4. Main potash, sodium and magnesium fertilisers

	Approximate nutrient content %
Muriate of potash, KCl	60 K_2O
Sulphate of potash, K_2SO_4	50 K_2O
Kainit	14–30 K_2O, 9–18 Na
Chilean potash nitrate	10–15 K_2O, 9–18 Na
Sylvinite, KCl.NaCl	18–22 K_2O, 25–28 Na
Salt, NaCl	37 Na
Nitrate of soda, $NaNO_3$	26 Na
Calcined magnesite, MgO	48 Mg
Kieserite, $MgSO_4.H_2O$	16 Mg
Magnesium kainit	at least 3.6 Mg
Epsom salts, $MgSO_4.7H_2O$	10 Mg

Plate 9.2. Cleveland potash mine   *Cleveland Potash Ltd*

Kainit (14–30 per cent K₂O; 9–18 per cent Na)

Kainit (14–30 per cent K_2O; 9–18 per cent Na)
Kainit is a naturally occurring mixed salt containing sodium and potassium chlorides plus magnesium sulphate. The name kainit as used in the UK refers to materials containing less than 3.6 per cent Mg. For magnesium kainit, see magnesium fertilisers later in this chapter. The major use for kainit is on sugar beet and other sodium responsive crops.

Chilean Potash Nitrate (15 per cent N; 10–15 per cent K_2O; 9–18 per cent Na)
A mixture of potassium and sodium nitrates, which is generally a relatively expensive source of nutrients.

Sylvinite (18–22 per cent K_2O; 25–28 per cent Na)
This is a naturally occurring mixture of potassium and sodium chlorides produced from the Cleveland mines (see Plate 9.2).

Agricultural Salt (37 per cent Na)
Agricultural salt is the major source of sodium used in the UK. Rock salt is supplied as a crystalline material from the Cheshire salt mines.

Nitrate of Soda (16 per cent N, 26 per cent Na)
This is a naturally occurring salt which is generally an expensive source of nutrients.

MAGNESIUM FERTILISERS

Magnesium may be applied in the form of magnesian limestone in the cases where lime is also needed. Otherwise it is usual to apply a specific magnesium fertiliser as listed in Table 9.4.

Magnesian Limestone
To be called a magnesian limestone for the purposes of the Fertiliser Regulations, a limestone must contain not less than 15 per cent MgO. A 5 t/ha application of an 18 per cent MgO limestone will apply 550 kg/ha of Mg. Where both lime and magnesium are required it is often economically worthwhile to transport magnesian limestone considerable distances, when compared with the cost of ordinary chalk or limestone plus a much more modest application of straight magnesium fertiliser. Another approach is to apply only part of the lime requirement as magnesian limestone. Sandy soils where magnesian limestone is the traditional liming material generally have a soil magnesium analysis index of 3, compared to 0 or 1 where little magnesium has been applied.

Experiments generally show magnesian limestone to be a slower acting source of magnesium compared to kieserite, but generally adequate if applied in the autumn before the responsive crop is grown. The long-term reserves from magnesian limestone are greater due to the much larger magnesium application rates normally applied.

Calcined Magnesite (48 per cent Mg)
Calcined magnesite (MgO) is the most concentrated source of magnesium fertiliser available. When applied to the soil it is converted to hydroxide and is faster in action than magnesian limestone. It is often sold in finely ground form which will increase its speed of action, but this is unlikely to be of practical significance for autumn applications. Some granular forms are made which disperse on wetting. Where low herbage magnesium is a problem, calcined magnesite can be used to dust the pasture and improve animal uptake of magnesium during grazing.

Work at Brooms Barn has shown that the temperature of calcining is important in determining the plant availability of the magnesium in calcined magnesite. Temperatures between 700–800°C are optimal. All sources used in the UK seem satisfactory from recent ADAS work, but the Fertiliser Regulations offer no protection.

Kieserite (16–17 per cent Mg)
Kieserite ($MgSO_4.H_2O$) is not easily soluble in water, but is readily available for plant uptake. It is the only material suited to seedbed application prior to growing a responsive crop.

Magnesium Kainit (at least 3.6 per cent Mg)
This is naturally mined kainit containing a varying but specified magnesium content.

Epsom Salts (10 per cent Mg)
Epsom salts ($MgSO_4.7H_2O$) are much more soluble and more expensive than kieserite. Their use is generally restricted to foliar sprays where their high solubility in water is an advantage.

TRACE ELEMENT FERTILISERS

Numerous products are available in the UK for both soil and foliar application. Many are in chelated forms, whilst others are simple inorganic salts. Chelates are stable molecules which bind the relevant ion and hold it in an available state even when applied to the soil. By contrast in many situations (see Chapter 8) simple inorganic salts are not effective when applied to the soil because the ion is quickly made unavailable for plant uptake due to adsorption by the soil. The main chelating agents used in fertilisers are listed below. They are referred to by their capital initial letters as shown.

EDTA–ethylene diamine tetra acetic acid
DTPA–diethylene triamine penta acetic acid
HEEDTA–hydroxyethyl ethylene diamine triacetic acid
EDDHA–ethylene diamine dihydroxy phenyl acetic acid.

EDTA is not generally suitable for soil applications. It is only stable in acid soils. The others are more stable in calcareous soils. EDDHA is the most effective soil applied iron chelate. The following sections are restricted to those elements and materials that are widely used under UK conditions.

Manganese
Manganese deficiency is only effectively treated by foliar sprays, with the exception of seed pelleting of sugar beet (see Chapter 17).

Manganese sulphate Mn $SO_4.H_2O$ (27–32 per cent Mn) is the cheapest material under most UK conditions. A large number of other commercial formulations of manganese are available. These include a range of liquid products based on both inorganic manganese (mainly sulphate) and chelated forms (mainly EDTA or DTPA).

Iron

Chelated products are expensive but necessary for the effective treatment of iron deficiency in perennial crops. Annual crops do not normally need treatment. For foliar sprays, EDTA-based materials are recommended; for soil application EDDHA has been shown to be the most effective.

Copper

For long-term control on severely deficient fields, copper sulphate ($CuSO_4.7H_2O$) crystals are generally recommended. For foliar application a wide range of materials, chelated and non-chelated, are effective as long as foliar scorch is avoided. The traditional copper fungicide materials based on copper oxychloride or cuprous oxide are more likely to scorch and are often little cheaper than EDTA chelated products.

Boron

Boron is not available in chelated form. The main fertiliser products used in the UK are borax and Solubor. Borax is sodium tetraborate, $Na_2B_4O_7.10H_2O$, which is moderately soluble in water. Various sodium tetraborates are used in boronated compound fertilisers. Solubor, approximate composition $Na_2B_8O_{13}.4H_2O$, is highly soluble and is the main material used for specific boron application to soil or growing crops. In recent years, a number of liquid formulations of boron have been introduced on the UK market.

SLOW RELEASE FERTILISERS

Slow release fertilisers are invariably more expensive than normal inorganic materials and are rarely justified for use on field-grown crops. Their main use is for container-grown horticultural crops. Slow release is achieved in two main ways. These are either slow breakdown in the soil by chemical or biological processes or by diffusion through some form of coating of the fertiliser granules.

Materials available include urea-formaldehyde (N), IBDU (iso butylidene di urea) (N), sulphur-coated urea (N), resin-coated fertilisers (N or NPK) and magnesium ammonium phosphates (N,P,Mg).

Chapter 10

COMMERCIAL PRODUCTS

THE RANGE of fertiliser materials discussed in Chapter 9 is commercially available to the farmer in a wide variety of products as straights and compounds, solids and liquids. Some of the main types of product on the market are considered in this chapter along with aspects of their storage, handling, application and agronomic effectiveness.

THE FERTILISERS REGULATIONS, 1977

The Fertilisers Regulations 1977 are published as Statutory Instrument No 1489 and cover the composition and labelling of fertilisers sold in Great Britain. The regulations cover the designation of solid fertilisers as EEC fertilisers which may then be traded under this common standard with any other EEC country. The non-EEC part of the regulations relates to liming materials, liquid fertilisers and other materials containing plant nutrients sold in the UK but less commonly sold between countries.

The regulations cover primarily:
1. Use and meaning of prescribed names and descriptions of materials.
2. Limits of variation of the nutrient content.
3. Manner of marking and labelling materials.

Nitrogen in EEC fertilisers must be declared as total N and also the content of:

- nitric (nitrate) nitrogen
- ammoniacal (ammonium) nitrogen
- ureic nitrogen
- cyanamide nitrogen.

If a product is non-EEC designated, only total N and ureic N if greater than 10 per cent by weight are needed. Phosphorus is declared as P_2O_5

Table 10.1. EEC Statutory nitrogen fertiliser label

EEC fertiliser	
Ammonium nitrate	
Total nitrogen (N)	34.5%
Ammoniacal nitrogen	17.2%
Nitric nitrogen	17.3%

Table 10.2. EEC Statutory NPK fertiliser label

NPK fertiliser 13:13:20	
Nitrogen (N) Total nitrogen (N) of which	13.0%
Nitric nitrogen (N)	4.3%
Ammoniacal nitrogen (N)	8.7%
Phosphate pentoxide (P_2O_5) P_2O_5 soluble in neutral ammonium citrate and in water	13.0% (5.7%P)
of which water soluble P_2O_5	11.8% (5.1%P)
Potassium oxide (K_2O) water soluble K_2O	20.0% (16.6%K)

Table 10.3. EEC Statutory NPK fertiliser containing rock phosphate label

NPK fertiliser containing soft ground rock phosphate 20:10:10	
Nitrogen (N) total of which	20.0%
Nitric nitrogen (N)	10.0%
Ammoniacal nitrogen (N)	10.0%
Phosphorus pentoxide (P_2O_5) soluble in mineral acids of which	10.0% (4.4%P)
P_2O_5 soluble in neutral ammonium citrate and in water	9.0% (4.0%P)
P_2O_5 soluble in water	6.0% (2.6%P)
P_2O_5 soluble only in mineral acid	1.0% (0.4%P)
Potassium oxide (K_2O) soluble in water	10.0% (8.3%K)

with the equivalent P content in brackets. Total P_2O_5 is required and the percentage P_2O_5 soluble in a particular extractant is commonly necessary (see Table 9.3). Potassium is declared as K_2O with K in brackets and magnesium as MgO with Mg in brackets. Examples of current labelling are shown in Tables 10.1, 10.2 and 10.3. Details of the sampling and analysis legislation used in the enforcement of these regulations are contained in: Fertilisers (Sampling and Analysis) Regulations 1978, Statutory Instrument No 1108; Fertilisers (Sampling and Analysis) (Amendment) Regulations 1980, Statutory Instrument No 1130.

TYPES OF FERTILISER

Two main types of fertiliser are used, straights which contain one nutrient only and compounds which contain two or more nutrients required by the crop. The main straight fertiliser sold in Great Britain is ammonium nitrate (33.5–34.5 per cent N). Many of the other materials discussed in Chapter 9 are available as straights. The definition becomes less precise in the case of materials such as kainit which are not processed but contain more than one nutrient. Compounds, the majority of which contain NPK but sometimes NP, NK or PK only, are fertilisers which have been produced to contain varying proportions of two, three or sometimes four nutrients. Each of these main types of fertiliser is sold in a range of physical forms— granules, prills, solutions or suspensions—resulting in the wide range of individual products on the market. Some non-water soluble phosphates are formulated in granules which break down quickly when moistened, thus retaining the speed of action of dust-sized particles.

SOLID FERTILISERS

The majority of the UK market is supplied by solid fertilisers. Most of these products, whether containing one or more nutrients, are formulated as granules generally in the range 2–3 mm diameter. Very few fine materials are now sold due to their unsatisfactory spreading characteristics. Nearly all materials sold in the UK are of a very high standard of physical specification. The amount of fines is minimal and caking of the products during storage is much reduced compared even to ten years ago. This is due to improvements in the technology of coating, particularly the more deliquescent materials. Some of the naturally occurring salts, such as kainit, can suffer caking problems.

Granules
Strictly speaking, granule should be used to define a particle that has been formed from a wet slurry, generally passed through a rotary granulator. Chemically it may be one compound or a mixture of two or three ingredients. Granules are normally coated with oil and then clay to ensure a free-flowing, non-caking product with a good storage life.

Prills

A prilled fertiliser is produced by spraying a solution of the fertiliser into the top of a tall prilling tower in much the same way as lead shot is made. Most ammonium nitrate and urea is formulated in this solid form for use as straight fertiliser. Ammonium nitrate prills are usually 2–3 mm diameter. Urea is usually 1–2 mm diameter.

Compounds

There are two basic types of solid fertiliser which contain two or more nutrients. Many but not all of the compounds sold are formed by making a slurry of the ingredients and subsequently drying and granulating the final product. Each granule then contains each nutrient.

The alternative which has become more popular in recent years is bulk blending. This is a dry process in which the individual solid ingredients are mixed together as required to produce a specific fertiliser. The main problem is the risk of subsequent separation of the ingredients. This is minimised by using ingredients of similar particle size and density. Magnesium and sodium are commonly incorporated in bulk blends.

The main ingredients used in UK produced bulk blends are:

- ammonium nitrate
- diammonium phosphate
- triple superphosphate
- potassium chloride
- calcined magnesite
- rock salt.

Storage and Handling

Much solid fertiliser in the UK is handled in 50 kg sealed polythene bags. The system provides flexibility in the choice of grades, easy identification and a simple check on application rates. Their main disadvantages are the cost and physical effort of handling, especially on farms where large quantities are used.

In recent years palletisation of 50 kg bags (see Plate 10.1) or the use of 0.5, 0.75 and 1 tonne bags (see Plate 10.2) have been widely introduced, reducing handling problems. Much bulk blended fertiliser in arable areas of England is handled in large bags or in bulk. Few farms store solid fertiliser in bulk; most bulk delivered material is for immediate spreading. Maintaining dry storage conditions for bulk storage on farms is too risky for most farmers.

Plate 10.1. Handling 50 kg bags   *UKF Fertilisers Ltd*

Application

The main machines used for spreading solid fertiliser are either broadcasters or full width applicators. Broadcasters are based on one or two spinning discs or an oscillating spout. Full width applicators employ various metering devices. The wider bout machines which have become very popular in the last few years use a pneumatic spreading mechanism. These are popular for tramline-based cereal systems, matching the sprayer bout width.

Range of Solid Compounds

Each solid fertiliser manufacturer in the UK offers a range of compounds containing different ratios of nutrients. In some cases these are marketed for specific crops, soil types or areas of the country. Commercial needs dictate that there is considerable variation in the number and type of compounds marketed by each company. Most compound manufacturers produce most if not all of the popular nutrient ratios given in Table 10.4. The actual nutrient contents may vary slightly from the specific examples given. The number of nutrient ratios produced by individual manufacturers varies from perhaps ten to nearly forty, depending on marketing policy. The bulk blend manufacturers market a wide range of nil nitrogen compounds for autumn application, including magnesium, sodium and boron as appropriate.

Plate C5. Potassium deficiency in potatoes

Plate C6. Magnesium deficiency in potatoes

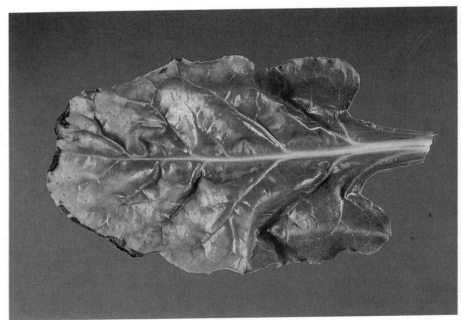

Plate C7. Magnesium deficiency in sugar beet

Plate C8. Magnesium deficiency in sprouts

Plate 10.2. Handling half tonne bags   *Norsk Hydro Fertilisers Ltd*

Table 10.4. Popular solid compound fertiliser ratios

% N	% P_2O_5	% K_2O
22	11	11
15	15	21
9	24	24
17	17	17
17	8	24
29	5	5
24	4	15
26	0	13
0	25	25

Plate 10.3. Liquid fertiliser application   *Hargreaves Fertiliser Industries Ltd*

LIQUID FERTILISERS

There are three main types of liquid fertilisers sold in the UK:

1. Clear solutions – non pressure.
2. Clear solutions – pressurised.
3. Suspensions.

Clear Solutions – Non Pressure

This is the most important type of liquid fertiliser sold in the UK. It may be broadcast (see Plate 10.3) or placed and needs neither injection nor sophisticated storage tanks. This type of fertiliser is sold as straight N usually 26 per cent N by weight (33 kg N per 100 l) or in NPK compounds. To achieve this straight N concentration, a mixture of ammonium nitrate and urea is used. Neither ingredient on its own can be formulated to this concentration.

Compound liquid fertilisers containing NPK, NP or NK are formulated using:

- ammonium nitrate
- urea
- ammonium phosphates

- ammonium polyphosphates
- potassium chloride.

Solubilities do not allow clear solution liquids to reach the concentrations of solid compounds. Due to the higher bulk density of liquids (1.2–1.3 g/l) compared to solids (0.9–1.0 g/l), equal quantities of nutrient whether solid or liquid have very similar volumes. The weight of liquid is invariably greater.

A source of confusion with liquid fertilisers is the expression of nutrient content. Some are sold on a volume basis (kg nutrient/100 l) and some on a weight basis (kg nutrient/tonne). Because volume varies with temperature, the Fertiliser Regulations require a weight-based nutrient content declaration as for solids. It also means that when comparing liquid and solid prices, care must be taken to convert to the same standard basis.

Storage and Handling
Some on-farm storage of liquids is necessary in most circumstances, but this can usually be limited to 50 per cent of annual usage. Storage tanks are expensive, and must be of suitable construction and sensibly sited so as to minimise the risk of pollution in the event of vandalism or accidental spillage. The great advantage of liquids is the ease of handling. They are easily pumped to fill the storage tanks and then into the applicator tank as required.

Application
Various techniques of application have been developed (see Plate 14.1). For application to bare soil or to grassland or young cereals where the risk of yield loss due to foliar scorch is low, spray application may be satisfactory. Liquid applicators normally use stream or flood jets. The larger droplet size is much less likely to scorch crops. Commonly the applicator is used for other agrochemicals after changing the jets. Problems with corrosion prevent the use of liquid fertilisers through ordinary sprayers. Scorch risk is virtually eliminated if a dribble bar system is used. Adaptations are available for both combine drilling and placement for potatoes.

Range of Clear Liquid Compounds
The main liquid fertiliser suppliers offer a range of about twelve different nutrient ratios in their list of compounds. The range of ratios is similar to the popular solid list but of necessity the nutrient contents are slightly lower.

Plate 10.4. Injecting aqueous ammonia on grassland   *Hargreaves Fertiliser Industries Ltd*

Clear Solutions – Pressurised

The two main pressurised clear solution fertilisers available on the UK market are anhydrous and aqueous ammonia. Very little anhydrous ammonia is currently used; limited quantities of aqueous ammonia are injected, mainly for grassland (see Plate 10.4) and brassica vegetables. The agronomic aspects of ammonia as a nitrogen fertiliser are discussed in Chapter 9.

Some aqueous ammonia is sold containing a nitrification inhibitor which slows the conversion in the soil to nitrate. As this is mainly used on grassland, it is discussed in Chapter 15. At present, there is little evidence of benefits for arable crops.

Handling, storage and application of ammonia require specialised equipment, so it is generally applied as a complete contract service. As application is slow compared to other materials, low application rates are relatively expensive.

Suspensions

Following development of suspension fertiliser technology in the USA, a limited amount is now made in England. Currently these are applied by a contract service and offer a limited range of NP and NPK materials. Overall nutrient concentration is comparable to solids. The compounds are made by mixing clear urea/ammonium nitrate solution with standard NP and NPK suspensions. These are made from ammonia, ammonium phosphate and potassium chloride suspended with finely divided clay.

The main limitation of suspensions is the need for regular agitation during storage and continuous agitation during application. Equipment is available to achieve satisfactory storage and application. The products do not have any agronomic advantages over solids or clear liquids.

FARM CHOICE

For most farms, agronomic efficiency is not a major factor in deciding between solids or clear liquids. The limitation of crop scorch from liquids can be overcome with suitable choice of application equipment. Either solids or liquids may offer specific crop advantages considered important on a particular farm, but these will generally be minor.

The major factors on which choice of fertiliser type will be based are farm organisation and economics. It is important to compare the complete costs of materials, storage, handling and application to reach a rational decision. In general in the UK, solids are cheaper to buy than liquids. However, liquids may be cheaper to handle and apply. Often the 200–400-ha arable farm with very limited labour can justify liquids, while solids are cheaper for the smaller and the larger farm.

COMPARISON OF NUTRIENT COSTS

Straight Fertilisers

To compare the relative costs of materials divide the cost of a bag by the number of kilograms (kg) of the nutrient contained. This gives the cost per kilogram of the nutrient for comparison with other materials containing the same nutrient.

Example: Compare superphosphate (18 per cent P_2O_5) costing £70/t with triple superphosphate (47 per cent P_2O_5) costing £141/t.

superphosphate   18 per cent = 18 kg/100 kg = 9 kg/50 kg bag

$$\text{cost/kg } P_2O_5 = \frac{70}{1000} \times \frac{50}{9} = 39p$$

triple superphosphate 47 per cent = 47 kg/100 kg = 23.5 kg/50 kg bag

$$\text{cost/kg } P_2O_5 = \frac{141}{1000} \times \frac{50}{23.5} = 30p$$

If comparing solids with liquids, be sure to work on weight-based nutrient contents for both.

Compound Fertilisers

Compounds can be rather more difficult to compare. In order to compare like with like, a notional cost of each product can be calculated using standard prices for each nutrient. For this example 36p, 28p and 15p per kg for N, P_2O_5 and K_2O respectively have been used. It is important that current figures are used.

By comparing the difference between the notional price and the actual price, the compound representing best value for money can be identified. The steps involved in this calculation are:

- Multiply the percentage of each nutrient by the appropriate standard price and add these together.
- Divide this sum by 10 to give the notional price (£/t).
- Divide the actual price by the notional price; the lowest value indicates the best value for money.

Example: Which of these three compounds represents best value?
A. 20:10:10 @ £127/tonne
B. 20:9:9 @ £121/tonne
C. 22:11:11 @ £137/tonne

Following the steps outlined above:

A. Notional price $= (20 \times 36 + 10 \times 28 + 10 \times 15) \div 10 = 115$

$$\frac{\text{Actual price}}{\text{Notional price}} = \frac{127}{115} = 1.10.$$

B. Notional price $= (20 \times 36 + 9 \times 28 + 9 \times 15) \div 10 = 110.7$

$$\frac{\text{Actual price}}{\text{Notional price}} = \frac{121}{110.7} = 1.09.$$

C. Notional price $= (22 \times 36 + 11 \times 28 + 11 \times 15) \div 10 = 126.5$

$$\frac{\text{Actual price}}{\text{Notional price}} = \frac{137}{126.5} = 1.08.$$

In this example compound C offers best value for money, although the differences are not substantial. This assumes that the compounds are comparable in terms of storage and spreading qualities.

Chapter 11

ORGANIC MANURES

THE TWO main organic manures applied to agricultural land are animal wastes and sewage sludges. Application of these manures achieves two purposes. It provides an effective means of disposal of materials that would be very expensive to process in other ways. It also allows the recycling of organic matter and nutrients back to the soil. While disposal and nutrient recycling are both achieved, there is often conflict over choice of rate and timing of application between cheap disposal and most effective nutrient use. A further constraint on disposal is the risk of pollution of streams or ground water if very high rates are applied at inappropriate times. Leaching through the soil can occur or in extreme cases direct run-off of manures into ditches or streams may take place. Odours are a problem in some situations.

The main consideration of this chapter is the contribution of the nutrients in organic manures to the immediate and longer term fertiliser needs of crops. This demands an understanding of the rate at which the total nutrient content of a manure becomes available for plant uptake from the soil. It is soon evident that the subject is far from precise. Manures are very variable products, often difficult to apply accurately and releasing nutrients in the soil at a rate very dependent on environmental conditions. As might be predicted, nitrogen utilisation presents more problems than phosphate or potash. Soil phosphate and potash levels have been built up on many farms in the UK by regular additions of farmyard manure for 100 years or more. It is very common to find that fields nearest the buildings are high in phosphate while off-lying fields are low. This reflects the more common grazing and farmyard manure application on the nearest fields. By contrast, the distant fields have been cut for hay and received less farmyard manure. These fields may well be very low in potash. This picture demonstrates that an important feature of regular manure use is the residual value of the nutrients applied over several years.

ANIMAL MANURES

A wide range of manures and slurries is produced by different cattle, pig and poultry production systems. These have varying nutrient values. Slurries are defined as faeces plus urine with a varying amount of added water from washing down or rainfall. The amount of dilution that has occurred due to added water is the most important factor determining nutrient content per unit volume. Manures are defined as

Table 11.1. Composition of farmyard manures and fresh, undiluted slurries (on fresh weight basis)

Type	Approximate dry matter %	Nitrogen % N	Phosphate % P_2O_5	Potash % K_2O	Magnesium % Mg
Farmyard manure					
Cattle	25	0.6	0.3	0.7	0.04
Pigs	25	0.6	0.6	0.4	0.04
Poultry:					
deep litter	70	1.7	1.8	1.3	0.40
broiler litter	70	2.4	2.2	1.4	0.22
in-house, air-dried droppings	70	4.2	2.8	1.9	0.40
Slurry (fresh and undiluted)					
Cattle	10	0.5	0.2	0.5	0.05
Pigs:					
dry meal fed	10	0.6	0.4	0.3	0.03
pipeline fed	6–10	0.5	0.2	0.2	0.03
whey fed	2–4	0.3	0.2	0.2	0.03
Poultry	25	1.4	1.1	0.6	0.12

Source: ADAS.

faeces plus urine absorbed in some bedding material, usually cereal straw or wood shavings. The relative proportion of straw and the degree of straw decomposition vary considerably. Due to all the variables involved, a single sample of a particular manure or slurry is not a reliable basis for assessing nutrient content in the laboratory, unless it is a particularly uniform product. An alternative policy is to use average figures for nutrient content as given by ADAS (Table 11.1). These assume no dilution of slurries and well-decomposed manures. For slurries it is necessary to reduce the nutrient values by the amount of dilution that has occurred.

Production and Usage

The main interest in this chapter is the animal manures from housed livestock. Nationally this may be taken as all manures from pigs and poultry, 50 per cent from cattle and none from sheep. The total quantities for the UK have been estimated at 300,000t for cattle (70 per cent slurry, 30 per cent farmyard manure), 70,000t for pigs (50 per cent slurry, 50 per cent farmyard manure) and 85,000t for poultry (90 per cent battery or broiler manure). Figures for the amounts of excreta

Table 11.2. Approximate amounts of excreta produced by livestock

Type of livestock	Body weight kg	Amount of excreta (faeces and urine or droppings) litre/day	Moisture content of excreta %
1 Dairy cow	500	41	87
1 Beef bullock	400	27	88
1 Pig – dry meal fed	50	4.0[a]	90
1 Pig – liquid fed (water:meal ratio 25:1)	50	4.0[a]	90
1 Pig – liquid fed (water:meal ratio 4:1)	50	7.0[a]	94
1 Pig – swill fed	50	14.0[a]	98
1 Pig – whey fed	50	14.0[a]	98
1 Dry sow	125	4.5	90
1 Sow + litter to 3 weeks	170	15.0	90
1,000 Laying hens	2,000	114	75
1,000 Broilers (+ litter)	—	68 kg	30

(a) Amounts of excreta produced over liveweight range 20–90 kg, ie. production per pig place.
Source: ADAS.

produced by housed livestock are given in Table 11.2. Figures from the *Survey of Fertiliser Practice* show how animal manure use in 1981 was split between the major crops (Table 11.3).

Where animal manures are produced and utilised on the farm, two possible approaches to efficient nutrient use may be adopted. Estimates of the nutrient content per tonne or 1,000 l can be used. A better alternative in situations where the necessary data are available is to work out the nutrients excreted by housed stock on the farm. This demands knowledge of the number of stock and the days housed. This is easily done for pigs and poultry. Similarly, it may be calculated for a dairy farm remembering that some slurry is produced during milking in the grazing season. The other essential information is knowledge of the

Table 11.3. Animal manure use in England and Wales, 1981

	Application % fields
Winter wheat	12
Winter barley	12
Spring barley	19
Maincrop potatoes	42
Sugar beet	30
Oilseed rape	5
Kale	43
2–7-year leys	41
All crops	17
All crops and grass	27

Source: *Survey of Fertiliser Practice.*

area over which the manure from a particular number of stock is spread. A particular advantage of this approach is that the actual weight or volume of manure or slurry produced does not need to be estimated.

Cattle and Pigs

Urine from these animals contains about half the total nitrogen excreted and 65 per cent of the potassium, but very little phosphorus. The urine nitrogen is mainly urea which is readily available for crop uptake from the soil. The faeces contain mainly undigested proteins. These decompose slowly in the soil so the nitrogen and the phosphorus are slowly available over a number of years. The potassium is virtually all available in the first year.

Experimental work on crop response to manures and slurries has shown very variable results even from carefully sampled and analysed materials. The figures given in Table 11.4 are the percentage available in the first year as recommended by ADAS. Of the 25 per cent nitrogen figure for cattle and pig farmyard manure, perhaps 10 per cent is present as readily available ammonium–N plus urea. In slurries, 50 per cent of the nitrogen is present as ammonium–N or urea. The available nitrogen as shown in Table 11.5 is mainly in this form. The annual release from the breakdown of organic constituents is fairly small. The amount of slurry nitrogen available to the next crop is much more variable than from farmyard manures. Application in autumn and early winter may result in loss by leaching of some of the urine nitrogen. Leaching will be greatest on spring-cropped sandy soils

**Table 11.4. Proportion of total nutrients available in the
season of application**

Type	N	P_2O_5	K_2O
		% available	
Farmyard manure			
Cattle	25	60	60
Pig	25	60	60
Poultry:			
deep litter	60	60	75
broiler litter	60	60	75
in-house air-dried	60	60	75
Slurry			
Cattle	30[(a)]	50	90
Pig	65	50	90
Poultry	65	50	90

(a) For slurry applied as a surface dressing to grassland. When incorporated into
   the soil soon after spreading the nitrogen may be 50 per cent available.
Source: ADAS.

receiving high winter rainfall. Whenever slurry is applied to the soil
surface, much of the urea is lost as ammonia to the atmosphere unless
it is rapidly cultivated into the soil. Similarly all the urine nitrogen can
be lost from bare soil in a dry spring. This leads to great variability in
results of experiments on its nitrogen fertiliser value.

The amounts of available phosphate and potash per tonne in Table
11.5 are those available in the first year. Over a period of years
virtually all the PK content will become available and contribute to the
soil analysis level. The figures in Table 11.6 are the ones to use if the
long-term maintenance of soil PK levels is the main concern, rather
than the first year crop response. As long as animal manures are
applied evenly and at a known rate, calculation of the phosphate and
potash applied for annual crop response or maintaining soil level can
be made. Plate 11.1 shows field application of farmyard manure.

It is interesting to consider the PK nutrient balance on an intensive
dairy farm. If a cow is housed for five months and produces about 40 l
of faeces plus urine per day, the total PK excreted is about 12 kg P_2O_5
and 31 kg K_2O. A field at soil index 2 for phosphate and potash, cut
three times for silage, requires 90 kg P_2O_5 and 250 kg K_2O/ha/year. If
the slurry from five cows is returned to 1 ha of silage ground every
year, it will need only 30 kg/ha P_2O_5 and 100 kg/ha K_2O per year of
fertiliser.

Pig slurry from pig fattening units will usually contain some copper

Table 11.5. Available nutrients in farmyard manure and slurries

Type	Available nutrients		
	N	P_2O_5	K_2O
Farmyard manure		kg/t	
Cattle	1.5	2.0	4.0
Pig	1.5	4.0	2.5
Poultry:			
deep litter	10.0	11.0	10.0
broiler litter	14.5	13.0	10.5
in-house, air-dried	25.0	17.0	14.0
Slurry (undiluted)	$kg/m^{3(a)}$ or kg/1000 litre[b]		
Cattle (10% dry matter)	1.5[c]	1.0	4.5
Pig (10% dry matter)	4.0	2.0	2.7
Poultry (25% dry matter)	9.1	5.5	5.4

(a) For slurry diluted 1:1 with water divide these figures by two. For slurry diluted 1:2 with water divide these figures by three.
(b) It is assumed that 1 litre of slurry weighs 1 kg.
(c) 2.5 kg/m³ if incorporated into the soil after application.
Source: ADAS.

which is fed to increase the rate of food conversion. Most is excreted, resulting in an undiluted slurry content of 0.05 kg Cu/1,000 l of slurry. This may cause toxicity to sheep if slurry-treated grass is grazed before the leaf is washed by rain. Toxicity to crops is unlikely until 280 kg/ha of copper has been applied. This is equivalent to 5.6 million l of slurry.

Table 11.6. Total nutrients in farmyard manure and slurries available over several years

Type	Total nutrients		
	P_2O_5	K_2O	Mg
Farmyard manure		kg/t	
Cattle	3	7	0.4
Pig	6	4	0.4
Poultry:			
deep litter	18	13	4.0
broiler litter	22	14	2.2
in-house, air-dried	28	19	4.0
Slurry (undiluted)	kg/m^3 or kg/1000 litre		
Cattle (10% dry matter)	2	5	0.5
Pig (10% dry matter)	4	3	0.3
Poultry (25% dry matter)	11	6	1.2

Source: ADAS.

Plate 11.1. Application of farmyard manure

Howard Rotavator Ltd

Poultry

Two manures are produced by the poultry industry, battery manure and broiler litter. The battery manure may be fresh or partially air-dried from deep pit houses. Broiler litter is generally based on wood shavings or straw. All poultry manures are high in nitrogen compared to pig and cattle products. Table 11.1 gives the average nutrient contents of the various types of manure. Due to its high nitrogen content, excessive applications commonly cause lodging in susceptible crops such as cereals.

About 60 per cent of the nitrogen in poultry droppings is present as uric acid. This readily breaks down in the soil to ammonium–N. A further 10 per cent is in the ammonium–N form with 30 per cent present as undigested food residues. Thus up to 70 per cent of the total nitrogen is readily available for crop uptake. Autumn application may result in some of the available nitrogen being lost by leaching. Ammonia loss is also considerable if battery manure is left on the soil surface during warm, dry weather. It is difficult to make good use of the nitrogen content.

Battery manures may contain up to 0.2 kg/t of zinc where zinc is used as a feed supplement. Zinc toxicity problems are unlikely below

560 kg/ha of zinc applied. This is equivalent to 2,800 t of battery manure.

Broiler manure also has a high content of available nitrogen. The total nitrogen content varies depending mainly on the proportion of litter to droppings. Percentage available nutrient figures are given in Table 11.2 and available nutrients content in Table 11.3. The total PK content for use in calculating maintenance applications is given in Table 11.4.

Recent Experimental Results
In the last few years ADAS has carried out a series of experiments on the nutrient value of cow slurry for silage production. The efficiency of crop use of slurry nitrogen and fertiliser nitrogen was compared on several sites over three years. Early spring applications of slurry were generally used about 30 per cent efficiently compared to ammonium nitrate fertiliser, but the variability in efficiency for the first cut was considerable. The figure of 50 per cent efficiency commonly used in the past was rarely achieved in practice. In these experiments slurry applied after the first cut in May–June was generally no more than 20 per cent efficient, compared to fertiliser nitrogen. Residual effects in the second cut after application were small for both timings. The target slurry application rates were 80 and 160 kg/ha nitrogen in these experiments.

Work at Gleadthorpe EHF has studied the long-term benefits of broiler litter, battery manure and farmyard manure applied annually to arable crops on sandland. Manures were applied in late autumn/winter and ploughed in. Highest annual yield was frequently obtained without extra fertiliser nitrogen. A longer term yield benefit has also built up on those plots receiving regular high rates of manures which could not be achieved by annual fertiliser nitrogen alone. Work by Rothamsted Experimental Station at Woburn, also on sandy soil, has shown similar residual benefits.

Storage of Animal Manures
In most situations, slurry or manure is not spread on the land daily. Much is stored on the farm for several months. Farmyard manure is generally spread once a year. When farmyard manure or broiler manure is stacked in the open, considerable nitrogen and potassium can be lost by leaching of rainwater through the heap. Some seepage losses of liquid from the heap may also occur. Gaseous loss of ammonia takes place when the heap is moved. The main loss occurs after spreading if the manure is not incorporated. The water content of the heap may change during storage.

The main loss from stored slurry is as gaseous ammonia; this is a particular problem of stored battery manure. Up to 20 per cent of the total nitrogen may be lost after a few months' storage. Agitation or aeration as part of the slurry treatment system will increase losses. However, the greatest nitrogen loss will usually occur after field application.

The costs of slurry storage are generally greater than any financial advantage from better nitrogen utilisation. Covered storage of farmyard manure is unlikely to recoup more than the cost of a polythene sheet in terms of better nutrient utilisation. When comparing the NPK values of animal manures with inorganic fertiliser, it is essential that the full transport and spreading costs are included to provide a meaningful comparison.

SEWAGE SLUDGES

Much sewage sludge produced in the UK is applied to agricultural land. In some situations application is done by the Water Authority; in others the sludge is handled by the farmer or a contractor on his behalf. Plate 11.2 shows machinery for injection of liquid sludge. Sewage sludges vary not only in nutrient content but also in their content of a wide range of other elements. Some of these are toxic to plants if high levels accumulate in the soil; others may be damaging to animals or humans eating the crops grown. For this reason guidelines for the maximum application of the various elements to agricultural land have been agreed for use in England and Wales. The Water Authorities work to these guidelines in applying their sludges and should provide an analytical service to monitor the levels in soils receiving regular sludge applications. As long as the guidelines are observed, problems are unlikely. However, the dangers should not be underestimated. Fields that will not grow crops or even weeds are not unknown in the UK. These have resulted mainly from regular disposal of large quantities of industrial contaminated sludge earlier this century, before the dangers were fully appreciated.

Nutrient Value

Some raw untreated liquid sludge is applied to agricultural land, but much is anaerobically digested at the sewage works to produce liquid digested sludge. This is a black liquid containing about 4 per cent dry matter and is spread by tanker or raingun. The other main sewage sludge is sludge cake containing 40–50 per cent dry matter. This is a solid which can be handled and spread by farmyard manure machinery.

Plate 11.2. Injection of liquid digested sewage sludge   *J. and H. Bunn Ltd*

Sewage sludges contain nitrogen and phosphorus in worthwhile amounts. Table 11.7 gives average values of nutrient content and the quantities available to crops in the year following application.

Liquid digested sludge contains most of its nitrogen as ammonium–N. If applied in the spring up to 60 per cent is available for crop growth. Untreated liquid sludge contains only 33 per cent available nitrogen. The phosphorus content of these sludges is about 50 per cent available in the first year. The figures in Table 11.7 should be doubled for maintenance situations although complete breakdown in the soil may take several years. Much of the available nitrogen is lost during dewatering to produce sludge cake, thus the availability figure is reduced to 15–20 per cent. Phosphorus availability remains the same as for liquid digested sludge. Injection of liquid digested sludge is carried out in some areas. This will not reduce the leaching risk, but reduces odour problems and ensures minimal nitrogen losses to the atmosphere.

Limitations to Use
Sewage sludges may contain levels of a number of elements which can adversely affect crops or human or animal health. For this reason, maximum levels for a number of elements have been agreed and these

Table 11.7. Typical nitrogen and phosphate content of sewage sludges (as spread)

Type of sludge	Dry matter (%)[b]	Nitrogen (N)		Phosphate (P_2O_5)	
		Total (%)	Available[a]	Total (%)	Available[a]
Liquid digested sludge	4	0.20	1.2 kg/m³	0.15	0.8 kg/m³
Liquid undigested sludge	5	0.18	0.6 kg/m³	0.15	0.8 kg/m³
Undigested cake	25	0.75	1.5 kg/t	0.65	3.3 kg/t
Digested cake	25	0.75	1.1 kg/t	0.90	4.5 kg/t

(a) The term 'available' in this table means available in the first cropping year.
(b) The 'per cent total solids' of a sludge may be quoted. This is the same as the 'per cent dry matter'.
Source: ADAS.

levels should not be exceeded. To ensure that soil levels on fields treated with sewage sludge are not exceeded, water authorities sample and analyse fields receiving sludge on a regular basis. It is advisable to apply only a proportion of the permitted dressing in any period of 10 years. Crop uptake of most of these elements is much reduced at high pH. Soil pH levels should be maintained at 6.5 or above on sludge-treated land.

To minimise the risk of animal disease, grazing intervals of at least three weeks for treated and six months for untreated sludge should be observed. Sludge is best avoided on land producing vegetables or fruit to be eaten raw. Untreated sludge should never be used. The risk of spread of potato cyst nematode means that use should be avoided on cyst-free land on which potatoes, bulbs or nursery stock may be grown. The only other crop problem that may occur is boron toxicity if high amounts are applied to sensitive crops such as cereals or potatoes. As long as application is not above 4.5 kg/ha boron in the first year, and 3.5 kg/ha in subsequent years, problems are unlikely.

The current maximum levels advised in England and Wales are given in Table 11.8. Any one of these elements may limit the sludge application to a particular field. Zinc is the commonest limiting element. The figures relate to the cultivated horizon or the top 7.5 cm if surface applications are made to permanent grass.

STRAW

In some circumstances cereal straw and other arable crop straws are chopped and incorporated into the soil rather than baled and carted or burnt. Most soils do not show any short- or long-term benefit from straw incorporation. The main difference between baled and burnt or

Table 11.8. Soil concentrations and limits of application of elements to safeguard human or animal health

Element	Soil concentration (mg/kg total element)		Limit of application[a] (kg/ha)
	Typical content of uncontaminated soil	Recommended maximum soil concentration	
Zinc	80	300	440
Copper	20	135	230
Nickel	25	75	100
Cadmium	0.5	3	5
Lead	50	250	400
Mercury	0.1	1	2
Molybdenum	1	4	6
Chromium	50	600	1,100
Selenium	0.5	3	5
Arsenic	10	20	20
Fluorine	200	500	600

(a) These rates assume soil initially uncontaminated and 20 cm depth of incorporation.
Source: ADAS.

incorporated is the 8 kg/t of K_2O contained in the straw. This is left in the ash after burning.

As straw has a wide C:N ratio due to a low nitrogen content, its initial breakdown takes up nitrogen from the soil. If this is likely to restrict crop growth, an extra 10 kg nitrogen per tonne of straw may be worth while to avoid competition for nitrogen. For autumn-sown cereals, most medium and heavy soils can provide adequate nitrogen for straw breakdown and crop uptake.

ORGANIC FERTILISERS

A number of organic fertilisers of animal or plant origin are marketed in the UK. Most are applied for their nitrogen content which is released following breakdown of the fertiliser by micro-organisms in the soil. They are generally expensive sources of nitrogen compared to inorganic forms and their nitrogen release is weather dependent. Many have traditionally been used for horticultural crops, but the market is diminishing. They include dried blood and shoddy.

Dried Blood (up to 13 per cent N)
A satisfactory nitrogen source that is nitrified fairly rapidly in the soil. Nearly all its nitrogen is readily available.

Shoddy (12–15 per cent N)
Shoddy is wool waste which contains 12–15 per cent nitrogen when pure. As long as it consists of wool and not synthetic fibres, it is a quick-acting nitrogen source. Various grades have traditionally been marketed, particularly for use on hops.

Organic Compounds
Pelleted fertiliser produced by an extrusion technique may be formulated to include organic nutrient sources. These are sometimes called semi-organic fertilisers. In these products, some of the nitrogen and phosphorus is present in organic form. Various sources such as composted wastes, dried poultry manure, dried sewage sludges and other organic wastes are used as the organic component in these products. The speed of nitrogen release will depend on the source used.

Chapter 12

SOIL AND PLANT ANALYSIS

SOIL ANALYSIS has been used for many decades in the UK as a basis for fertiliser recommendations and for diagnosis of nutrient deficiencies in crops. Plant analysis has been restricted mainly to the latter use but is used along with soil analysis in determining the fertiliser requirement of perennial crops.

Soils may be analysed for each nutrient in numerous different ways. The main interest is in methods of measuring the amount of a nutrient which correlates with crop performance. Methods which measure the amount of nutrient in available form can then be calibrated in terms of crop fertiliser requirement or the likelihood of nutrient deficiency at a particular soil level.

MAKING FERTILISER RECOMMENDATIONS

Soil analysis is routinely used to measure pH, lime requirement, available phosphorus, potassium and magnesium. On particular soil types, individual trace elements may be included. This is discussed for each trace element in Chapter 8. Specific aspects of pH and lime requirement are covered in Chapter 3. Numerous methods of soil analysis for nitrogen have been examined as discussed in Chapter 4, but none has made a major contribution to the improvement of nitrogen recommendations for field crops in the UK.

SOIL SAMPLING

Nutrients levels in field soils do not change rapidly provided a sensible fertiliser policy is being followed. The frequency with which analysis is worthwhile depends on the circumstances of soil type, cropping and previous nutrient problems. Every four to five years is a reasonable aim

Plate 12.1. Soil sampling of arable land

to keep a check on any changes that may be occurring. The better organised the farm fertiliser policy, the less the need for soil analysis to monitor changes. In well-organised situations every ten years will be adequate for P, K and Mg. But pH will need more constant attention on non-calcareous soils.

A soil analysis can only be as good as the care with which the field sampling is carried out. Firstly the appropriate sampling tool should be used. This will be a screw auger as in Plate 12.1 or cheese-corer-type auger for most arable and horticultural cropping situations where the sampling depth is 0–15 cm. If the land has been ploughed annually, sampling depth is not critical and a garden trowel can be used. However, if the field is not ploughed every year as for instance under apples or direct drilled wheat, it is essential that uniform cores to the correct depth are taken. This is because a gradient of nutrients will commonly have built up. It is particularly important to include the top few centimetres in each core.

For grassland a cheese-corer or tubular type of auger as in Plate 12.2 should be used. This will give the best chance of retaining the top few centimetres. Sampling depth for permanent grass and long leys is

Plate 12.2. Soil sampling of permanent grass

7.5 cm. Again nutrient gradients dictate the need for accurate sampling depth. If grassland is being ploughed out it is better to sample to 15 cm after ploughing. The special needs of soil sampling before planting fruit and hops are covered in Chapter 20.

A common problem is the inclusion of cores from atypical parts of the field. Avoid sites of old manure heaps and bonfires and leave out headlands and areas round trees, pylons, gateways and anywhere stock may congregate. If a sample is taken soon after lime or fertiliser application, the analysis will be higher than if left long enough for the fertiliser to react with the soil. Where lime or high rates of other nutrients are applied, it is best to wait twelve months if possible. Sampling just before or after harvest before fertiliser is applied for the next crop is generally the best choice. Where small amounts of phosphate or potash have been applied, a reasonably reliable figure will be found three months after application.

There can be no precise instruction on how to split up a field for sampling and how many hectares can be represented by one sample. If variation in pH is dealt with by indicator testing as recommended in Chapter 3, the area sampled can be large, up to 20 ha, as long as it has been cropped and fertilised uniformly for a long time and the soil textural variation is small. Some textural variation in a sampling unit may be acceptable if the field can only realistically be treated as one for fertiliser application. However, it must be remembered that where field boundaries have been removed the new field will often vary in nutrient levels depending on the previous treatment of the individual smaller fields. These differences can show even after 100 years or more. The field is usually the maximum size of sampling unit; soil

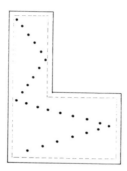

Figure 12.1. Pattern of field soil sampling

texture variation may justify splitting this into two or three. Only occasionally will it be appropriate to take a composite sample from more than one current field.

To represent an area of land, 20–25 individual cores should be taken and bulked together to give a single soil sample for analysis of a half to one kg in weight. These cores should be taken by walking the field in a W pattern and taking one core every 10–20 m depending on field size. The field should be traversed as shown in Figure 12.1. Try to avoid walking in the line of normal lime and fertiliser spreading operations on the field. Samples should always be put into clean polythene bags to avoid contamination and labelled as soon as taken, preferably with prepared labels. Do not take half a dozen samples and then try to remember which came from which field.

Nutrient Gradients

Under a regular ploughing policy the topsoil layer is well mixed and phosphorus, potassium and organic matter are uniformly distributed throughout the soil. By contrast, when the soil is not inverted by ploughing, a gradient in the content of some nutrients builds up over a period of years. This is most noticeable in arable crops under very shallow cultivation or direct drilling (Table 12.1). Similar effects occur under long-term grass or fruit crops.

The gradient of nutrients depends on the mobility of the individual nutrient when added to the surface of non-ploughed land. The least mobile major nutrient is phosphorus. A high surface content builds up due to fertiliser use and return of crop residues (Table 12.1). The lower soil layers are slowly depleted due to crop uptake. For this

Table 12.1. Nutrient gradients under different cultivation systems (Grange Piece, Boxworth EHF, after eight years)

	Depth cm	pH	P mg/l	K mg/l
Direct drilled	0–2.5	6.0	66	409
	2.5–5	6.9	56	390
	5–7.5	7.3	21	324
	7.5–10	7.3	15	280
	10–20	7.3	14	264
Shallow cultivation	0–2.5	6.9	60	428
	2.5–5	7.1	57	365
	5–7.5	7.3	43	309
	7.5–10	7.4	24	299
	10–20	7.5	16	280
Ploughed	0–2.5	7.7	23	336
	2.5–5	7.7	24	339
	5–7.5	7.6	24	320
	7.5–10	7.7	22	295
	10–20	7.7	26	301

Source: ADAS.

reason non-ploughing should only be adopted on soils in P index 1 or above. If the soil is P index 0 the phosphate level in the plough layer should be built up first. If the soil is index 1, occasional ploughing is desirable to prevent depletion of the lower topsoil layer. For perennial crops, phosphate levels should be raised before establishment if practicable. Potassium moves rather more rapidly than phosphorus. Although a gradient will generally occur in non-ploughed fields it is unlikely to limit yield or require special attention (Table 12.1). A gradient will build up even if potash fertiliser is not applied due to the return of crop residues or animal manures to the soil surface.

SOIL ANALYSIS

Soil analysis services are available from ADAS, most commercial fertiliser companies and commercial laboratories not involved in the fertiliser industry. Most laboratories in England and Wales now use the ADAS analytical methods for phosphorus, potassium and magnesium. The main exception is the Norsk Hydro method for available phosphorus. Some smaller companies use non-ADAS methods particularly for trace elements.

As methods of measuring available nutrients are empirical, they are only of value if they can be shown to correlate with crop uptake and yield response on UK soils. The main advantage of using common methods is that the field experimental information for the interpretation of the figures on numerous soils and crops is extensive. For this reason it is desirable that the current ADAS methods for phosphorus, potassium and magnesium, which have been used since January 1971 in England and Wales, should not be changed without very strong reasons.

Phosphorus

The Olsen method of available phosphorus has become very popular worldwide. The current ADAS method is based on this technique. Phosphorus is extracted from soil at $20°C \pm 1°C$ with sodium bicarbonate buffered at pH 8.5. Control of the temperature and pH levels is very important in this technique. Soil is shaken at a ratio of 1 vol soil:20 vol extractant for thirty minutes and the phosphorus is measured colorimetrically after developing a blue colour by reaction with ammonium molybdate. Full details of this and other ADAS

Table 12.2. ADAS nutrient indices for phosphorus, potassium and magnesium

Index	Phosphorus (mg/l)	Potassium (mg/l)	Magnesium (mg/l)
0	0–9	0–60	0–25
1	10–15	61–120	26–50
2	16–25	121–240	51–100
3	26–45	241–400	101–175
4	46–70	401–600	176–250
5	71–100	601–900	251–350
6	101–140	901–1,500	351–600
7	141–200	1,501–2,400	601–1,000
8	201–280	2,401–3,600	1,001–1,500
9	over 280	over 3,600	over 1,500

analytical methods are published in MAFF, 'The Analysis of Agricultural Materials', *RB. 427* (3rd edn.).

The results are expressed in mg/l P and also provided in an index form for ease of use (Table 12.2). Interpretation of the index system is covered in Chapter 13. The ADAS method for phosphate is less satisfactory on acid soils, but this is not a major problem except in upland areas. Of more concern is its poor ability to measure the residual value of water-insoluble phosphates used on grassland. There may be a case for the use of an acidic extractant on permanent grass where insoluble phosphate fertilisers have been used.

Norsk Hydro use a resin method for extraction of available phosphorus. This gives very similar results to the ADAS method when expressed on an index basis, but removes a greater amount of phosphorus from the sample.

Potassium

Potassium is readily measured by soil analysis and a range of extractants may be used which will give similar answers. The current ADAS method uses molar (M) ammonium nitrate with an extraction ratio of 1 vol soil : 5 vol extractant. The shaking time is thirty minutes and potassium is read directly using a flame photometer. Analytical results are expressed in mg/l K and also in index form (Table 12.2).

Various methods of measuring the long-term potassium supplying power of a soil have been studied, in particular by Rothamsted and by Norsk Hydro. It is a subject of debate as to whether a knowledge of soil type and clay mineralogy is adequate to predict soil K reserve or whether analysis is better. One problem is the subsoil K contribution on soils with high K reserves which is not normally sampled or measured by analysis.

Magnesium

Magnesium, like potassium, presents relatively few problems of assessing the amount available for crop uptake. The current ADAS method of extraction is the same as for potassium. The two elements are measured on the same extract. Magnesium is determined using an atomic absorption spectrophotometer. Analytical results are expressed in mg/l Mg and on an index basis (Table 12.2).

Other Nutrients

ADAS methods of analysis for available nutrients in soils are outlined in Table 12.3.

Table 12.3. ADAS methods of analysis for available nutrient levels in soils

Nutrient	Method of extraction
Boron	boiling water for 5 mins
Cobalt	0.5 M acetic acid
Copper	0.05 M ammonium EDTA at pH 7.0
Magnesium	M ammonium nitrate
Molybdenum	acid ammonium oxalate
Phosphorus	sodium bicarbonate pH 8.5 at 20°C
Potassium	M ammonium nitrate
Sodium	M ammonium nitrate
Zinc	0.5 M acetic acid

Source: Analysis of Agricultural Materials ADAS RB 427 HMSO.

Soil Testing Kits

In recent years a number of DIY soil testing kits have been marketed in the UK. Indicator testing for pH is usually part of these kits and is generally satisfactory as long as an appropriate indicator mixture is supplied. The main problem of any DIY kit use is obtaining a representative sample for analysis. The test is carried out on a teaspoon full of soil which must represent 5 or 10 ha. In a soil analysis laboratory, samples are dried, ground to 2 mm and mixed before subsampling for analysis. It is difficult to subsample accurately while the sample is moist.

If sampling can be carried out satisfactorily, there is no reason why a simple test kit should not give a rough estimate of P or K level. It will be difficult to relate this reliably to the ADAS index system without comparing results with a routine soil testing laboratory. For most farmers and growers, their skill is best used in taking good samples of fields in which they know the detail of variation in soil type and previous use. Analysis is better left to a fully equipped laboratory with constant quality control checking of the results produced. Some kits can produce reliable results in the hands of knowledgeable enthusiasts.

NUTRIENT LEVELS IN UK SOILS

Over the past thirty years nutrient levels, particularly of phosphorus, have increased in arable and intensive grassland areas. An analysis of data compiled by Bunns shows that many East Anglian soils are index 3 for phosphate with over 20 per cent index 4 or above (Table 12.4).

Table 12.4. Phosphorus, potassium, magnesium and pH levels in arable East Anglian soils—13,335 samples, 1980–83

pH	Less than 6.5 9.8%		6.6–7.0 13.4%	Greater than 7.0 76.8%
P	Index 0 3.2%	Index 1 10.0%	Index 2 24.8%	Index 3 + 62.0%
K	Index 0 5.3%	Index 1 39.5%	Index 2 + 55.2%	
Mg	Index 0 7.8%	Index 1 48.6%	Index 2 + 43.6%	

Source: J & H Bunn Ltd, Great Yarmouth.

Table 12.5. ADAS representative soil sampling scheme 1979–83—percentage of samples at different nutrient indices in England and Wales

	Nutrient index					
	0	1	2	3	4	5+
Phosphorus						
Arable	3	10	26	36	17	8
Grass	15	26	26	23	7	2
Potassium						
Arable	2	20	52	19	5	1
Grass	6	40	43	8	2	0
Magnesium						
Arable	3	21	35	16	10	15
Grass	0	5	38	32	14	11

Figures from the ADAS Representative Soil Sampling Scheme which monitored nutrient levels and changes throughout England and Wales emphasise the differences between arable and permanent grass as shown in Table 12.5. The fields in this study are a randomly selected sample of agricultural fields in England and Wales. Comparison of the Bunns data with the national arable figures shows similar P and lower K and Mg figures in East Anglia, which is mainly due to the presence of a much higher proportion of sandy soils in the Bunns sampling area.

PLANT ANALYSIS

The use of analysis of plant material for predicting fertiliser requirement is limited to perennial crops in the UK. The interpretation is often associated with crop quality rather than just yield response. Grass may be sampled and analysed for various reasons. Samples are taken following the same principles as for soil samples. Soil contamination must be avoided. At each sampling point a small quantity of grass should be cut, not pulled by hand. These are bulked to provide a field sample. The other major factor is the stage of growth for sampling. The main reason for sampling is to assess the nutritional quality of the herbage for grazing animals and decide on the need for extra feeding or fertilising. For this the sample should be taken just before grazing.

Analysis for major nutrients and trace elements can be carried out as required using total methods of analysis. Interpretation depends on the digestibility of the herbage and the type of stock and is beyond the scope of this book. Analysis is generally restricted to farms with a known history of animal nutritional problems. Soil analysis is generally less satisfactory than direct analysis of the herbage in sorting out these problems.

Analysis of grass is also useful in some situations to help assess fertiliser need. Potash uptake at grazing or cutting stage can be monitored by grass potassium analysis. This can help assess the potash fertiliser requirement. Analysis of second cut silage for N:S ratio provides a reliable assessment of likely response to sulphur fertiliser on a particular field.

Leaf analysis for top fruit and both leaf and fruit analysis for apples are used to modify fertiliser use, particularly of nitrogen and potassium. The long-term storage quality of apples is influenced by nutrient content of the fruit. Fruit analysis is used as an indicator of storage potential and this may be improved over a number of years by modifying fertiliser practice. Leaf analysis is used mainly to give an indication of nitrogen uptake but is also useful for other nutrients. In apples it often correlates with fruit analysis. As several seasonal factors influence plant nutrient levels, data for two or three years from the same orchard are necessary before major changes in fertiliser use are made. Leaf manganese levels give an early warning of soil acidity problems. Details of sampling and interpretation are given in Chapter 20.

DIAGNOSIS OF CROP DEFICIENCIES

When sampling and analysis are carried out to decide whether a crop is suffering from a nutrient deficiency, different principles and methods are used. In some situations the whole field crop may be uniform, but visually poor in growth. More often, poor areas of growth are obvious in contrast to better parts of the field. It is generally better to take both soil and plant samples if possible. This will allow a more complete diagnosis.

Soil Sampling
If the whole field is to be sampled, follow the procedure described earlier. To represent poor patches, the soil within a number of patches should be sampled to the appropriate depth. A good soil sample from outside the patches, but no further away than necessary, should also be taken. The main principle to follow is to take a representative sample of the problem areas, not the whole field. If acidity is a possible cause of the poor growth, this should be checked with indicator during sampling. Relying on the analysis of soil from around the roots of a plant sample is generally less representative and therefore a less satisfactory basis for diagnosis, particularly of acidity or any nutrient where a gradient of concentration with depth is likely.

Plant Sampling
Plant analysis is widely used for the diagnosis of major nutrient and some trace element deficiencies. As for soil, the samples should fully represent the good and poor areas. It is important that enough total plant material is provided for analysis. Where possible whole plants, including roots, should be taken. The appropriate parts of the plant can then be subsampled for analysis in the laboratory. Where it is not possible to take whole plants, the general principle for leaf sampling is to take 100 just fully expanded leaves. In some cases there may be justification for taking leaves showing symptoms, but if these are older leaves, as is often the case, diagnosis and treatment can be more difficult to resolve. The nutrient content of the younger leaves may be satisfactory. The methods of leaf sampling for fruit are given in Chapter 20.

Analysis and Diagnosis
Soil analysis is carried out following the available methods discussed earlier. Nitrate–N may be useful in giving an indication of the topsoil nitrogen status, particularly if the plant is young with a small root system. Plant analysis is carried out using methods which determine total content, expressed as mg/kg of dry matter. If only one nutrient is low, deficiency of that nutrient in the plant is confirmed, particularly if it matches the visual symptoms. Diagnosis is more difficult if more than one nutrient is low. Low manganese with low nitrogen does not confirm manganese deficiency. If all the major nutrients are low, root

function is in question, rather than a particular deficiency. This is likely if soil analysis shows satisfactory levels as commonly happens. Magnesium deficiency is often induced when root activity is poor. It is rare to find a genuine soil-related deficiency of P, K or Mg unless the soil index is 0. Potatoes will show phosphorus and potassium deficiencies at index 1.

Chapter 13

BASIS FOR RECOMMENDATIONS

WHILE NUMEROUS factors influence fertiliser recommendations for individual crops as discussed in succeeding chapters, common broad principles apply. In this chapter the basis for decisions on the quantities of nitrogen, phosphorus, potassium and magnesium that should be applied is covered.

Two principles can be distinguished. Crop response is the primary consideration. Where field experiments show an annual economic yield increase, this will dictate fertiliser policy. This applies particularly to nitrogen for non-leguminous crops and NPK for potatoes. The second consideration is the maintenance of satisfactory soil nutrient levels so that maximum yield can be achieved with appropriate fresh fertiliser. This applies particularly to phosphorus and to a lesser extent for potassium and magnesium. Soil analysis provides the means of monitoring maintenance of soil PK and Mg levels.

RESPONSE CURVES

When a field experiment is carried out comparing a number of different application rates of a nutrient, a response curve of the results relating yield or some other parameter to the rate of nutrient applied can be drawn. A typical nitrogen response curve is shown in Figure 13.1. To produce a complete curve the range of levels tested must start at zero and extend above the level at which no further yield increase occurs. The increment between the rates applied should not be greater than 50 kg/ha. To obtain an accurate but complete curve, at least six rates of nutrient are needed. The greater the number of points, the more precise the curve and the more accurate are the conclusions derived from it. Five is the minimum number of nitrogen levels needed in any experiment designed to answer the question, 'what is the economic rate to apply in these circumstances?' The success of the

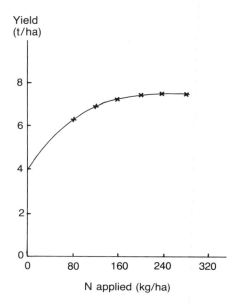

Figure 13.1. Nitrogen response curve for winter wheat
Source: ADAS

experimental design will depend on whether the choice of rates spans the economic optimum. The incremental difference should still not exceed 50 kg/ha.

The fitting of a response curve to a set of data is a complicated statistical exercise and a source of much debate, which is beyond the scope of this book. In general when fitting mathematical curves to biological data, it is more reasonable to use the model curve which best fits the single set of data, rather than constrain data from a large number of experiments to one type of curve. There is no right answer in the sense of one type of curve representing the relationship between yield and rate of applied nutrient to a particular crop.

The rate of response is often calculated from a nutrient response curve; this is the slope of the curve. It must be clearly defined, however, which part of the curve is being used: sometimes the initial slope is assessed; often the rate of response up to the optimum is quoted, which will generally be a lower rate because it includes the flatter part of the curve as it nears the optimum. The other main pitfall of rate of response comparisons is that the data may not always include a nil application rate.

Economic Optimum

The advantage of fitting an appropriate mathematical curve to a particular set of data is that a precise economic optimum can be derived and the statistical error of that optimum can be measured. The economic optimum is defined as the point on the curve at which the increase in value of crop is equal to the increase in cost of nutrient. Figure 13.2 gives an example of this by replotting the data in Figure 13.1 on a monetary basis.

Most recommendations are derived from a simple cost comparison relationship. For example if wheat is valued at £120/t and nitrogen costs 36p/kg, the ratio of kg grain needed to balance the cost of nitrogen is 3 kg grain : 1 kg N. The same principles may be applied to other nutrients and crops. Depending on the shape of the response curve, the optimum may be well defined or, as often happens when the soil supply of the nutrient is high, very poorly defined. The shape of the curve above the optimum is also important. If the curve declines above the optimum, the risk of yield loss from applying too little or too much fertiliser is about equal. More often the slope of the curve flattens but

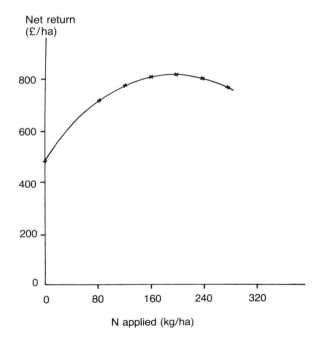

Figure 13.2. Economic value of nitrogen response on winter wheat
Source: ADAS

continues gently upwards. In this circumstance, the risks are not equal. Applying too much fertiliser will only slightly reduce the economic benefit, whereas too little fertiliser may give a considerable economic penalty.

If a major change in the value of the crop or cost of the fertiliser occurs, the effect on the economic optimum again depends on the shape of the curve. For nitrogen, the steep response curve found for cereals means that the economic optimum will only fall by about 10 kg/ha N if the ratio increases from 3:1 to 4:1. On the other hand, if the response curve is shallow as often occurs for phosphate on potatoes, the economic optimum may be halved by a widening of the ratio.

The nitrogen response curve also allows questions concerning the effects of limitation on nitrogen use to be answered. For cereals, a 50 per cent reduction in nitrogen use below the economic optimum will result in about a 10 per cent reduction in yield. For grass, the comparable figure is a 20 per cent yield reduction. These figures only relate to the first-year effect. Reduced nitrogen fertiliser use will decrease the soil nitrogen supply over a number of years and thus the actual economic

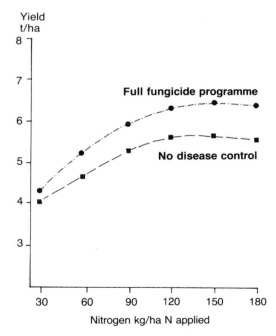

Figure 13.3. Nitrogen response of winter barley with and without a fungicide programme
Source: ADAS

optimum nitrogen fertiliser rate will increase. If a reduced rate of nitrogen use is enforced for several years, the yield reduction will increase above the annual figures quoted above.

Interactions

Many experimental designs examine more than one input. Work on potatoes often examines N, P and K and work on cereals often includes nitrogen and fungicide or growth regulator. The yield and shape of the nitrogen response curve is commonly influenced by the presence and amount of the other variable. Figure 13.3 shows the nitrogen response of winter barley with and without a fungicide programme. Yield is reduced when the other input is omitted and the economic optimum is lower. A true interaction only occurs if the increase in yield from the two variables in combination is greater than the sum of the increases from the two variables independently.

In this example, yield is less and the total nitrogen uptake by the crop is also less where lack of fungicide is restricting yield. However, the percentage efficiency of fertiliser nitrogen use is unlikely to be changed. So the economic optimum nitrogen rate is reduced. Other factors such as root diseases or water stress affect crop root activity. These will reduce yield and often reduce the efficiency of fertiliser nitrogen use as well. In these situations the economic optimum may be the same or greater, while yield is actually reduced.

NITROGEN

Literally hundreds of individual nitrogen response experiments on a very wide range of crops have been carried out in the UK in the last thirty years. Since soil analysis has not provided the basis for grouping fields, other ways of classifying fields have been sought. The general

Table 13.1. Median and range of nitrogen optima for winter wheat on clay soils

	Mean yield at optimum N t/ha	Median N optimum kg/ha	Range of optima kg/ha
N index 0 (18 sites)	7.99	180	120–280
N index 1 (18 sites)	9.61	160	80–240

Source: ADAS.

approach has been to look at data from a number of experiments and relate the mean or the median economic optima to a chosen site factor such as soil type or previous cropping. While these can account for some of the variability in optima, the range is often still ± 50 per cent of the median optimum (Table 13.1). Separation on a soil type basis often does no more than separate peats from mineral soils. Previous

Table 13.2. ADAS N index system

Nitrogen index based on last crop grown		
Nitrogen index 0	Nitrogen index 1	Nitrogen index 2
Cereals	Beans	Any crop in field receiving large frequent dressings of farmyard manure or slurry
Forage crops removed	Oilseed rape	
Maize	Peas	
Leys (1–2 year), cut only	Potatoes	
Leys (1–2 year), grazed or cut and grazed, low N[a]	Forage crops grazed	Lucerne
	Leys (1–2 year), grazed or cut and grazed, high N[b]	Long leys, grazed or cut and grazed, high N[b]
Permanent pasture, poor quality, matted	Long leys, cut only	Permanent pasture, cut only, grazed or cut and grazed
Vegetables receiving less than 200 kg/ha N	Long leys, grazed or cut and grazed, low N[a]	
Sugar beet	Vegetables receiving more than 200 kg/ha N	

	Nitrogen index following lucerne, long leys and permanent pasture				
Crop	1st crop	2nd crop	3rd crop	4th crop	5th crop
Lucerne	2	2	1	0	0
Long leys, cut only	1	1	0	0	0
Long leys, grazed or cut and grazed, low N[a]	1	1	0	0	0
Long leys, grazed or cut and grazed, high N[b]	2	2	1	0	0
Permanent pasture, poor quality, matted	0	0	0	0	0
Permanent pasture, cut only	2	2	1	1	0
Permanent pasture, grazed or cut and grazed, low N[a]	2	2	1	1	0
Permanent pasture, grazed or cut and grazed, high N[b]	2	2	1	1	1

(a) Low N = less than 250 kg/ha N per year and low clover content.
(b) High N = more than 250 kg/ha N per year or high clover content.

cropping is the most useful variable and forms a major part of the ADAS recommendations system. The N index system (Table 13.2) classifies the field on a three class scale of index 0, 1 or 2 depending on the previous crop or, where grass is in the rotation, several years' previous cropping. In deriving this system, ADAS looked at numerous other field factors and soil analysis techniques but rejected them because they did not appreciably improve the prediction of the optimum. While individual sets of UK and European data support the use of various parameters, such as soil mineral nitrogen in the profile in early spring, features of winter rainfall or level of yield, none has been adequate on its own to provide an improved basis for nitrogen use on cereals or any other crop over the wide range of conditions in the UK.

The main hope for future progress is to use computer modelling to integrate all these factors rather than to examine them individually as has been done in the past. To do this, data on various aspects of crop nitrogen use which can only be achieved using N–15, as described in Chapter 4, are essential. These data are now becoming available from research under UK conditions. To improve the current position the following components need to be quantified:

- crop uptake,
- soil N supply,
- efficiency of fertiliser N use,
- fertiliser nitrogen losses from soil.

The amount taken up by the crop will inevitably involve working to a predicted yield. Work with N–15 allows the soil N supply to be measured in the crop. Refinement of this factor alone should allow improvement of the N index, which mainly reflects differences in soil N supply. The understanding of the turnover and mineralisation of crop and fertiliser residues has improved considerably in recent years. Efficiency of fertiliser use can be assessed using N–15, but will remain a major cause of variation in optima. Numerous factors which affect the crop after fertiliser application can cause a change in the efficiency of use.

PHOSPHORUS AND POTASSIUM

Similar principles of interpretation of response curve data apply for P and K as for nitrogen. The main difference between these nutrients and nitrogen is the much reduced risk of nutrient loss and the much smaller rates of response unless the soil is index 0 for either nutrient. Large yield increases from annual applications of phosphate or potash are less common than they were thirty years ago. Applications of fertilisers and organic manures over many years have built up the soil levels of phosphorus and potassium on many farms. In many cases these nutrients are now applied to maintain a satisfactory soil level rather than to give an immediate yield benefit. Soil analysis provides the means of deciding which policy of phosphate and potash use is appropriate on a particular field.

Table 13.3. Maintenance soil levels of phosphorus and potassium

	Phosphate	Potash
Potatoes/vegetables	index 3 (26–45 mg/l P)	upper index 2 (181–240 mg/l K)
Other arable and forage	index 2 (16–25 mg/l P)	lower index 2 (120–180 mg/l K)

Note: soil levels measured by ADAS methods.

Yield Response and Maintenance of Soil Reserves

Phosphorus and potassium are required in relatively large quantities by all crops. These requirements can be met from two sources; the soil reserves or fertilisers applied for the crop. The first reason for using P or K is to obtain an economic yield increase. The higher the soil nutrient level, the lower the probable yield response to fresh P or K fertiliser. Maximum crop yield will only be achieved when soil levels of phosphorus have been built up so that the yield response to fresh P fertiliser is small. If the soil phosphate level is index 0, most crops will not give maximum yield however much fresh fertiliser P is applied. Until this nutrient has been thoroughly incorporated into the topsoil roots are not able to take it up efficiently (see Figure 5.2 in Chapter 5).

As potassium is more mobile than phosphorus in the soil, fresh potash fertiliser can largely make up for low soil K levels but this does not always happen, particularly in dry seasons and on poorly structured soils. The aim should be to apply enough fertiliser to reach the guideline levels shown in Table 13.3. Yield responses are infrequent and often uneconomic when soil nutrient levels are at or above these figures, except for potatoes. Potatoes will often give small but economically worthwhile yield increases to fresh fertiliser at higher soil levels, but these levels need not be built up and maintained.

Soil Analysis
Soil analysis provides a measure of the amount of available P or K in a field sample at a particular time. Changes in soil levels of P or K are normally only seen after a number of years unless extremely high or low P or K rates have been applied. It usually takes at least three years to move one index up or down the scale, although K levels on sandy soils can change more quickly. Due to the variation of soils within a field, a comprehensive sampling technique is essential to provide a representative index level for a field. Duplicate samples will commonly vary by up to 5 mg/l P and 30 mg/l K. Poorly taken samples can vary by much more.

Fertiliser Policy
Depending on the soil level of P and K and the crop there are four policy options for a particular field:
1. Apply for crop response only (which will also build up the soil level if the amount applied is greater than the amount removed in the crop).
2. Build up the soil level (by applying more than will be removed in the crop).
3. Maintain the soil level (by applying the amount which will be removed in the crop).
4. Allow the soil level to run down (by applying less than the amount which will be removed in the crop).

The quantities of fertiliser applied should be enough to give the full yield response for the crop if the soil level of P or K is below the levels suggested in Table 13.3. More generous rates should be applied to build up the soil to these levels where financially possible. Where soil P levels are below these guidelines for a particular rotation they should be built up as soon as practicable, to ensure that yields are not limited in the future. This demands annual applications of at least 50 kg/ha P_2O_5 greater than offtake.

Soil K levels should be built up to the appropriate guidelines if possible, but sandy soils may not be able to maintain these levels. For these, an annual application determined by crop response is the only option available. When the guideline levels have been reached, a maintenance policy can be followed. Potatoes respond above the guideline levels and should continue to receive appropriate recommended rates. These can contribute to P and K requirements of other crops in the rotation, because the amounts applied are usually greater than those removed in the crop, especially of P. Where soil P or K levels are above the suggested guidelines they can be maintained or allowed to run down depending on personal choice and financial circumstances. When levels are above P index 4 or K index 3, full maintenance applications are difficult to justify.

Fertiliser Rates

The first step in deciding on fertiliser rates is to assess the behaviour of P and K in the soil. This information is available from twenty-five ADAS long-term experiments plus data from studies by Rothamsted Experimental Station, the earliest starting on Agdell field at Rothamsted in 1848 and still continuing. Where there is a need to build up the soil P level, this can be achieved on all mineral soils as long as more phosphate is applied than removed. To increase the soil by 10 mg/l P, experiments show that 400–600 kg/ha P_2O_5 extra fertiliser in addition to offtake is needed regardless of soil type. Much more may be needed if the initial level is below 10 mg/l on clay soils. The extra phosphate is best applied over a period of years rather than in one application. For phosphate balancing, P offtake by fertiliser addition is adequate to maintain the soil level once the required guideline levels have been reached. On sands the level may rise and on clays it may fall slightly, but the change over ten years will be small and is of little practical significance.

Experiments give a range of values for the amount of K fertiliser needed to raise the soil K in the plough layer by 50 mg/l. On medium soils, 400–600 kg/ha K_2O is needed, in addition to offtake. Most medium and heavy soils release some potassium each year from natural weathering of clay minerals. The Rothamsted experiments have shown that this release is substantial on the Saxmundham chalky boulder clay and only now shows signs of diminishing after eighty years' cropping without K fertilisers. On the clay with flints at Rothamsted, a somewhat lower release has continued for well over 130 years.

In working out the potassium maintenance requirements, an annual rate of release of 50 kg/ha K_2O per year may be assumed for chalky boulder clay or Lias clay soils. Less is known of the K release from other clay soils of different origins.

Sandy Soils

Sandy soils have a limited capacity to retain potash due to their low clay content. Thus it is not practicable to achieve and maintain high potash indices on these soils. For loamy sands, 150 mg/l is generally possible, but any attempt to achieve a higher level will result in considerable potash moving into the subsoil and some being lost by leaching. For sands, the practicable limit is no higher than 100 mg/l.

Table 13.4. Phosphate and potash content of crops

Crop	% DM (at harvest)	% P (DM)	% K (DM)	kg/t fresh P_2O_5	K_2O
Cereal – grain	85	0.4	0.55	7.8	5.6
– straw	85	0.08	0.8	1.6	8.2
Sugar beet – roots	22	0.15	0.8	0.8	2.1
– tops	16	0.3	3.0	1.1	5.8
Potato – tubers	22	0.2	2.2	1.0	5.8
Field beans	85	0.55	1.2	11	12
Dried peas	85	0.45	1.0	8.8	10
Oilseed rape	92	0.75	1.0	16	11
Grass – silage	20	0.3	2.0	1.4	4.8
– hay	85	0.3	1.8	5.9	18
Kale	15	0.35	2.8	1.2	5
Swedes	10	0.3	2.0	0.7	2.4
Maize – forage	25	0.2	1.2	1.1	3.6
Vining peas	18	0.4	1.5	1.7	3.2
French beans	10	0.45	2.0	1.0	2.4
Broad beans	15	0.45	2.0	1.6	3.6
Sprout – buttons	15	0.75	3.5	2.6	6.3
– stems	20	0.45	3.0	2.1	7.2
Cabbage	10	0.4	3.0	0.9	3.6
Cauliflower	10	0.6	4.0	1.4	4.8
Onions – bulb	10	0.3	1.5	0.7	1.8
Carrots	10	0.3	2.5	0.7	3.0
Redbeet	15	0.3	2.5	1.0	4.5
Bulbs	35	0.3	1.5	2.4	6.3

Source: ADAS.

The risk of K loss by leaching is considerable on these soils. Crop response is the main criterion for potash use on sandy soils. Annual applications are invariably needed.

Assessing Maintenance Requirements
Apart from allowing for K release on clay soils, the main factor determining maintenance requirements is the amount removed in the crop. Table 13.4 gives average amounts of P_2O_5 and K_2O in crops. The variation in P_2O_5 is small; K_2O contents vary much more.

Only in special circumstances is it likely to be worthwhile to analyse an individual crop to refine these figures. Using these figures and field yields, the amount of P and K removed each year can be calculated. A record of all P and K applied is also essential. For organic manures, the total PK content rather than the available content should be used in the maintenance calculation. Where crop residues are burnt in the field, it

Table 13.5. Phosphate and potash removed in average yielding crops

	Fresh yield t/ha	P_2O_5 kg/ha	K_2O kg/ha
Cereal – straw burnt or ploughed in	7 (grain)	55	39
– straw removed	7 (grain)	63	79
Sugar beet – tops ploughed in	40 (roots)	32	84
– tops removed	40 (roots)	63	246
Potato*	40	40	232
Field beans*	3	33	36
Dried peas*	3	26	30
Oilseed rape*	3	48	33
Grass – silage	50	70	240
– hay	5	30	90
Kale	60	72	300
Swedes	40	28	96
Maize – forage	35	39	126
Vining peas*	5	9	16
French beans*	8	8	19
Broad beans*	3	5	11
Sprouts – buttons	12	31	76
– stems removed	12	56	162
Cabbage	30	27	108
Cauliflower	30	42	144
Onions – bulb	30	21	54
Carrots	35	25	105
Redbeet	35	35	158
Bulbs	10	24	63

* haulm not removed. Source: ADAS.

Table 13.6. Forward planning of PK needed for the rotation

Field A – medium texture—no K release
– soil P, 25 mg/l (index 2)
– soil K, 220 mg/l (index 2)

Policy – to maintain current soil levels

Rotation	Yield estimate t/ha	Nutrient removal kg/ha P_2O_5	kg/ha K_2O
Sugar beet (tops ploughed in)	40	32	84
Barley (straw baled)	6 (grain)	47	34
	4 (straw)	6	33
Dried peas (haulm burnt)	3	26	30
Wheat (straw burnt)	8	62	45
TOTAL		173	226

Note: The total figures are the amount of P and K needed for the whole four-year rotation either as a single application before the sugar beet or split in some other way.

Table 13.7. Checking back on PK use over a rotation

Field C – clay texture – 50 kg/ha K_2O released per year
– soil P, 25 mg/l (index 2)
– soil K, 180 mg/l (index 2)
Policy – to maintain soil levels

Rotation	Actual yield	Nutrients removed kg/ha P_2O_5	kg/ha K_2O	Nutrients applied kg/ha P_2O_5	kg/ha K_2O
Oilseed rape	3	48	33	40	0
Winter wheat	8	62	45	60	0
Winter wheat	7	55	39	60	0
Winter barley	6	47	34	60	0
TOTAL		212	151	220	0

Note: Phosphate in balance. Potash removal less than soil release of 200 kg/ha over four years so well in balance.

may be assumed that little of the P or K in the residues is lost. From these data, the P and K balance (P and K applied minus P and K removed) can be calculated. If the balance is positive, soil reserves will be increased. If the balance is negative, soil reserves will be depleted.

Table 13.5 gives offtake data based on average yields. From this table it can be seen that a good crop of winter wheat removes more P per hectare than does a crop of potatoes. Cereals normally receive much less P than potatoes which explains why many fields growing good crops of cereals show little increase in their P index values.

Balance Sheets

Simple balance sheets of nutrient application and removal provide the basis for maintaining P and K levels on a field by field basis. Forward planning of application rates must obviously be based on expected yields. An example is given in Table 13.6. The alternative approach is to look back over a number of years to check whether P and K applications have been too low or too high. Assuming the soil was known to be at or above the guideline at the start of this period, inputs and offtakes can be added up to check whether the soil level has been maintained. If rates have been too high or too low, future rates can be modified accordingly. An example is given in Table 13.7. Where a maintenance policy is being used with reasonably accurate yield data, soil analysis for PK is only occasionally necessary to check that no unforeseen change has occurred.

Magnesium

All except sandy soils can generally be maintained in index 1 or above with appropriate fertiliser or from natural weathering of soil minerals plus atmospheric deposition. Information on long-term changes in soil magnesium levels is sparse. The levels of available magnesium are naturally very high (index 5 and 6) in some soil parent materials, notably Keuper Marl and marine alluvium. The main concern for fertiliser use is those relatively few soils which cannot maintain an adequate available Mg level (index 1 for most crops).

The soil level in most sandy soils will decline unless magnesium fertilisers are used regularly. Many sandy boulder clays and shallow chalk soils of around 20 per cent clay content can maintain index 1 without the need for fertiliser application. It is rare that magnesium fertiliser need be applied to any soil other than sandy textured topsoils, except for fruit crops.

If the soil analysis level is 15 mg/l or below (low index 0) on sandy soils, magnesium should be applied for all crops until the soil level is raised. Between 16 and 30 mg/l Mg (high index 0 to low index 1) an insurance application is recommended for responsive crops. This will achieve any yield response and maintain the soil level. Application is not worthwhile for cereals above 15 mg/l Mg. Offtake for most crops is in the range 5–10 kg/ha Mg. Magnesium should not be applied more frequently than every fourth year, as the residual value of the fertiliser will last for three years. If magnesian limestone is used as the liming material, magnesium levels of 100 mg/l or more (index 2–3) will be maintained even on sandy soils and specific magnesium fertiliser is not required.

Recommendations Using Compound Fertilisers

In practice many fertiliser recommendations for individual nutrients are applied as compound fertilisers. It is unlikely that a particular NPK fertiliser recommendation will be matched precisely by a commercially available compound. Where this problem occurs, the first priority is to apply the correct nitrogen rate. If this results in more phosphorus or potassium than recommended being applied, it may be possible to balance this by applying less in the next year. At worst the extra will help to build up the soil level. If the soil level of P or K is low, it is important not to apply less of this nutrient than recommended. A commonly used approach is to choose a moderate nitrogen content compound of the correct P:K ratio and then use straight nitrogen to make up the remainder of the nitrogen requirement.

Chapter 14

METHODS AND TIMING OF APPLICATION

WHILE METHODS and timing of fertiliser applications are specific for individual crops and are discussed in succeeding chapters, common principles are involved. Method of fertiliser application depends on the physical and chemical form of the product being applied. Some machinery is specific to one type of fertiliser on one crop such as placement of liquid fertiliser for potatoes. The majority of broadcast application machinery, whether for solids or liquids, can be used on a wide range of crops. Timing of application is constrained by several factors if the most efficient use of the nutrients is to be achieved. Problems of crop damage whether above or below the soil surface can be caused by several nutrients. Most timing decisions relate to the use of nitrogen where problems of denitrification and leaching of the nutrient may occur.

METHODS OF APPLICATION

Overall Broadcast
Most solids and liquids are applied broadcast on bare soil or over the growing crop. Both specific foliar feeding and the risk of crop scorch are discussed later in the chapter. The range of broadcast application machines is summarised in Chapter 10. For most applications, ordinary farm tractors are satisfactory. Only when the soil is wet and the bearing capacity is too low for a conventional tractor are other alternatives sought. Very steep slopes may present problems to grassland improvement. The limited situations in the UK where the farm tractor is inadequate have traditionally been met by aerial application. This is the best technique where slope is a problem or where crop damage from field application needs to be avoided. The success of aerial application depends both on operator skill and the field being free of obstacles.

In the last few years, low ground pressure vehicles in a multitude of shapes and sizes have taken over the job of early spring nitrogen topdressing. With these machines, application is possible virtually any day of the year as long as the level of crop damage is acceptable. Wheel ruts are no longer a problem.

A broadcast application can be applied evenly or unevenly. The following list of machine and operator faults is responsible for poor spreading:

- poor balance of spread
- inaccurate bout width
- incorrect operating height
- incorrect pto speed
- too high wind speed
- poor machine maintenance.

Different types of machine are susceptible to different problems. Wind and balance of spread problems are greater with spinners. Liquid sprayers and pneumatics both need regular maintenance of nozzles and spreading plates respectively. Accurate tramlining of combinable crops has provided a major advance in bout width accuracy on many farms in the last few years. ADAS surveys have shown that the physical condition of the fertiliser is rarely responsible for poor spreading.

Placement

A wide range of machines has been produced for the controlled placement of fertiliser materials. These include machines for side-banding applications to the soil surface to avoid crop damage and for placement of the fertiliser in the soil for greater efficiency of use.

Only a small amount of sidebanding where the fertiliser is applied to a restricted part of the soil surface is carried out in the UK. Most machinery has been farmer adapted. For sugar beet, deflector plates are advocated for keeping nitrogen applied at or soon after drilling off the line of the row, to avoid crop damage. A limited amount of sidebanding is also carried out, to apply solid fertiliser to crops which might be scorched. Sprouts are commonly topdressed in this way. Some machines have been modified by top-fruit growers to apply fertiliser to the herbicide strip in which the trees are growing and to avoid the grass alley between the rows. By this means fertiliser is only applied to the main crop rooting zone.

Fertiliser may be placed in the soil for two main reasons; to be more efficiently used by the young plant or to prevent loss of nutrient by volatilisation to the atmosphere. Combine drilling of cereals and placement of fertiliser for potatoes are the two main examples of the first type of placement. Each will be discussed in the appropriate crop chapter. The major advantage is the more effective use of phosphate on low phosphate soils. Response to placed nitrogen is also shown in some experiments. The advantages of placement are usually of doubtful economic value on soils of adequate PK index particularly if it incurs a delay in date of drilling or planting.

Work in the late 1970s at various Agriculture and Food Research Council (AFRC) Institutes showed yield benefits from the deep placement of phosphate and potash. Widespread testing of these findings by ADAS on commercial farm sites has shown few yield responses. In general, deep placement of phosphate and potash cannot be recommended on the basis of this work.

Some nitrogen sources can only be used effectively if placed in the soil. The main product in this category is liquid ammonia, either aqueous or anhydrous. For effective use, the injection should be carried out with no smell of ammonia. For anhydrous 15 cm depth is generally needed and for aqueous 10 cm is usually adequate except on sandy soils.

FOLIAR FEEDING

Foliar application can be an effective remedy for a crop suffering from a nutrient deficiency. Foliar feeding aims to wet the leaf and encourage the plant to take up the nutrients through its leaves. The principles involved are different from those for applying liquid fertilisers when the aim is to avoid wetting the crop foliage and invoking a risk of scorch.

Foliar feeding is most effective for trace element deficiencies. It is difficult for a crop to take up a substantial amount of a particular nutrient from one application. Benefits from major nutrients are small unless several applications are made. Magnesium deficiency can be effectively treated if three or four applications are applied. For trace elements a single spray is often adequate to provide the much smaller crop demand. Foliar spraying invariably carries a risk of crop scorch which limits the concentration at which most nutrients should be applied (Table 14.1). Recommendations for fruit crops are given in Chapter 20. To be effective, the foliar spray should be applied to a good leaf cover. To enter the leaf, the nutrients must pass through either the stomata or the epidermal cuticle. Mass movement through the stomata does not generally occur. Young leaves are generally more effective in taking in nutrients. The amount of leaf wax largely determines how much is taken up. Some crops such as onion and the upper surfaces of some leaves are difficult to permeate.

Adequate temperature and high relative humidity can increase uptake. Uptake is usually poor if the leaf is under water stress. To reduce the risk of scorch, nutrients should not be applied during conditions causing rapid evaporation. This leads to localised drops of high salt concentration on the leaf. Nitrates and chlorides are generally most damaging and best avoided. Non-ionic sources such as urea or chelated forms are less likely to cause scorch. A wetter should be included to ensure uniform application; most applications are ineffective without a wetter. The concentration of nutrient should not be too high but care must be taken to ensure that the spray does contain enough of the particular nutrient needed by the crop.

Table 14.1. Recommendations for foliar nutrient applications to arable and vegetable crops

	Rate kg/ha	Application rate l/ha
Urea	25	500
Potassium sulphate	20	500
Magnesium sulphate	20	500
Iron chelate (EDTA)	0.5	500
Manganese sulphate	8	250
Boron (Solubor)	5	250
Copper oxychloride	2	500
Sodium molybdate	0.25	500

Note: All foliar nutrient sprays should contain a wetter.

Wetters
A wetter or surfactant should be added when foliar feeding with an inorganic nutrient source, unless the product is formulated for foliar feeding or the nutrient is tank mixed with another product containing adequate wetter. In the latter situation, the mix should be checked by dipping a leaf in the tank to ensure that full wetting is achieved. When a wetter is needed, a non-ionic type is generally preferable. The following are listed in the 1985 *Approved Chemicals* booklet (Agricultural Chemicals Approval Scheme, MAFF Ref Book 380) and should be used at the rate recommended on the label.

Agral (Plant Protection)
Farmon Blue (Farm Protection)
PBI Spreader (Pan Britannica).

TIMING OF APPLICATION

Crop Establishment

A young plant may be damaged if a high osmotic pressure resulting from a high concentration of soluble fertiliser salts occurs in the soil water. Damage results from the physical disruption of cells and occurs to sensitive root or shoot tissue after germination. Seeds can usually take in enough water to germinate even in a high osmotic pressure environment. Problems only occur when the soil is relatively dry so that the osmotic concentration is high, because all the fertiliser is dissolved in the small amount of soil water. This is most likely to occur near the soil surface. The main problems result from soluble ionic materials notably ammonium nitrate, potassium chloride and sodium chloride applied in high concentration to the seedbed. Reduction of plant population is of greater economic significance in crops such as sugar beet or some vegetables which are drilled to a stand and demand that at least 70 per cent of the seeds produce a harvested plant. Problems occur but are of less significance in cereals and other crops with a higher seed rate relative to the minimum acceptable plant population.

Practical problems are greatest on sandy soils and almost unknown on peats. Most problems can be avoided by appropriate timing of nutrient applications. This will generally demand separating the nitrogen from the application of the other nutrients. The main problems occur when too much fertiliser is applied just before or after drilling. Combine drilling of cereals usually delays emergence by a few days due to osmotic effects slowing early growth but rarely reduces yield. Peas can suffer greater damage if combine drilled. If fertiliser is placed in contact with potato seed, considerable root and stem damage may occur, reducing the number of viable stems in some cases. Problems can usually be avoided if a limited amount of fertiliser is placed close to but not in contact with the seed tuber itself.

Crop Scorch

Damage to most crops can occur when they are topdressed with nitrogen in either solid or liquid form. Unlike foliar feeding the aim is to apply the fertiliser to the soil with minimal effect on the foliage. Problems with solids are generally few and restricted to the lodging of prills within the crop. They then take up a small amount of moisture and produce a necrotic lesion due to a localised high ionic concentration. Crops which have a cup shape where the leaves extend from the stem are most susceptible. Maize and tulips are notable for this characteristic and should not be topdressed. Crops vary in their sensitivity to damage and the economic consequences. Brassicas, particularly oilseed rape, are often spotted by fertiliser, but the only economic effect seen is when damage occurs to the growing point of particularly cauliflowers or sprouts. Any crop may be more seriously scorched if a batch of fertiliser contains appreciable dust size material which sticks to leaf surfaces.

Liquid nitrogen is potentially more damaging. Even with stream or flood jets and a large drop size, topdressing of vegetable crops should be avoided as only slight scorch could make the crop unmarketable. Dribble bar application is the safest technique. The range of liquid fertiliser application methods is shown in Plate 14.1. For most other crops, stream or flood jets will minimise damage and are acceptable. Normal fine spray jets should not be used. Winter oilseed rape has been found susceptible to scorch, particularly if topdressed with liquid during frosty weather in early spring. If appropriate equipment is not available for a crop, scorch risk can be minimised by application of liquid nitrogen while it is raining, so no concentrated liquid remains on the crop.

Nitrogen

Nitrogen fertiliser can be lost by leaching or by volatilisation to the atmosphere. Timing of application must take these risks into account. Problems of adverse effects on plant population have been mentioned earlier and will be detailed for specific crops in succeeding chapters.

The other major factor to take into account is soil dryness. If the crop has already built up an appreciable soil moisture deficit before the main application of nitrogen as may occur under winter cereals, the rate of uptake will be slow even if temperature and water supply for the crop are not limiting growth. For most efficient uptake the nitrogen needs to be in the zone of soil from which water is being taken up. This is a particular problem of application to grass after cutting or grazing in the drier parts of England. Frequent applications of nitrogen are most effective where no appreciable soil moisture deficit occurs. In some situations, particularly in the dryer parts of the country this factor should be taken into account in deciding on the appropriate timing of topdressing. Topdressing of winter wheat should usually be completed by the end of the first week in May. If the soil moisture deficit is greater than 25 mm, this date may be too late in a season when May rainfall is low. A consideration of both soil and crop factors is needed to arrive at the best nitrogen timing for a given set of circumstances. The aim must always be to reduce the risk of both loss and inefficient crop uptake.

SPRAYING. Fast surface application of liquid fertiliser before planting or for top dressing grassland and early cereals. Although specially designed for liquid fertiliser, the applicator may, with just a change of jets, be used for agricultural chemicals.

STREAM JETS. Nitrogen liquid fertiliser may be applied with stream jets when top dressing to produce larger droplets for better foliar run-off and reduced crop scorch.

DRIBBLE BAR. Where foliar protection is particularly important, such as in broad-leaved crops and later in cereals, liquid fertiliser may be dribbled through tubes below the crop canopy.

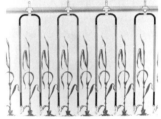

COMBINE DRILLING. Drills may be converted for liquid fertiliser where there are yield benefits on low fertility soils whilst still maintaining a high drilling rate. The drill hopper is kept solely for seed and corrosion is minimised by ducting the liquid fertiliser in separate tubes.

PLACEMENT. Liquid fertiliser can be injected through narrow tines for simultaneous placement in bands below and to the side of the seed at planting time. Used mainly for potatoes, but in certain crops such as brassicas side band placement post-planting may also be of benefit.

Plate 14.1. Application methods for liquid fertiliser   *J. W. Chafer Ltd*

Leaching Problems

Practical problems of nitrogen application in the spring are limited to sandy and shallow soils. For spring-sown crops, application policy should ensure that only part of the crop requirement is applied before the end of March on these soils. For autumn-sown crops a proportion of any autumn-applied nitrogen can be lost by leaching on most soils, especially if applied after crop growth has slowed down. Nitrogen applied after October 1st will be inefficiently used in most circumstances.

Topping up with nitrogen in the spring in addition to the normal application is rarely justified unless all the nitrogen for spring-sown crops on sandy soils has been applied earlier than recommended. Table 14.2 was formulated by ADAS Cambridge to provide advice based on the NVRS leaching model. Consideration is given to likely economic

Table 14.2. Recommendations for extra nitrogen top dressing

	Sandy soils		Medium and heavy soils	
	Excess rainfall			
Crop	50 mm	100 mm	50 mm	100 mm
Winter cereals oilseed rape, grass	no extra	no extra	no extra	no extra
Spring cereals	no extra	40 kg/ha N[a]	no extra	no extra
Sugar beet, onions	no extra	40 kg/ha N[a]	no extra	no extra
Potatoes,[a] brassicas	40 kg/ha N	80 kg/ha N[b]	no extra	40 kg/ha
Carrots parsnips	no extra	no extra	no extra	no extra

(a) Only if all intended nitrogen applied before rain,
(b) 40 kg/ha N adequate if half nitrogen intended to go on at tuber initiation.

return and any penalty associated with excessive use of nitrogen for particular crops. These recommendations are for extra nitrogen assuming the recommended rate had already been applied.

Phosphate and Potash

Where a yield response is anticipated, phosphate or potash should be applied annually. For phosphate all the fertiliser should be applied in the seedbed. For potash, autumn applications ploughed down for spring-sown crops are acceptable unless the soil level is below 100 mg/l when a spring application is necessary. For potatoes, at least some of the potash should be spring applied. For cereals at index 0 for phosphate or potash, combine drilling of the appropriate nutrient is worthwhile.

Where the soil level is being maintained at or above the guideline figures (see Table 13.3), the appropriate amount of phosphate or potash may be applied every one, two or three years in autumn or spring, before or after ploughing or drilling except for potash on sandy soils. Potatoes which respond above the guideline level should always receive at least some of the rotational requirement. Because potatoes respond to much more phosphate fertiliser than is removed in the tubers, a considerable residue is generally available for following crops. Otherwise the timing and method of PK application is very flexible. The main condition is that the full maintenance application is applied during the rotation.

All sandy soils should receive annual rather than infrequent potash applications. These will generally be applied for crop response rather than maintenance. On sands, potash is better applied after ploughing in late winter or spring to reduce the risk of leaching. Care should be taken to avoid high potash applications in the seedbed before drilling or planting. Failure to do so under dry conditions can reduce emergence or cause root damage.

Chapter 15

GRASSLAND AND FORAGE CROPS

UNLIKE THE crops in succeeding chapters, grass and forage crops must be considered in terms of animal utilised economic benefit rather than the field yield of a specific part of the plant sold at a price per tonne. Grass cut for hay or silage and harvested forage crops can be considered in terms of yield per hectare with an assumed value per tonne of a particular feeding quality. This allows a realistic calculation of economic fertiliser recommendations to be made. Grazed grass is the most difficult crop to assess. Many experiments have measured grass yield response by cutting at grazing frequency and relating grass dry matter production to nitrogen application, producing a response curve. Unfortunately the difference between this simple experimental approach and the production of milk or meat per hectare measured against nitrogen input is considerable. The latter has been attempted in relatively few experiments.

A further complication that must be taken into account is the range of species and situations included in the general description of grass and the wide range of soil/climate situations in which they are grown. In this chapter lowland grass (including fenced upland pastures with predominantly ryegrass swards) is separated from the hill and upland grazing situations where production is more severely limited by soil/climate. Latitude, aspect, soil type and sward management also interact, giving a very wide range of situations.

Another factor on animal production farms is the need to account for the return of nutrients to the fields either as slurry or farmyard manure or as faeces and urine during grazing. An assessment of economic fertiliser use must include the contribution from animal manures. The value of slurry and farmyard manure should be calculated as discussed in Chapter 11. The benefits of the faeces and urine from grazing animals are more difficult to assess but need to be included. Not least of the problems is the very uneven pattern of application, particularly under less intensive stocking rates. Evenness of application improves as stocking rate increases.

Grass grown for herbage seed is dealt with in Chapter 18. Many aspects such as establishment and partial use for grazing are common to forage grass discussed in this chapter.

LOWLAND GRASS

Most lowland grass swards are based on ryegrasses with a varying proportion of white clover. Other species, particularly cocksfoot or timothy, may be present but make little difference to the sward fertiliser requirement. The sward may be a sown ley in rotation with other crops, a sown long-term ley reseeded when necessary or a permanent pasture. Again differences in fertiliser requirement are small. Much depends on the intensity with which the sward is utilised.

Liming and pH
Most grass species will produce maximum yield at a pH as low as 5.5 on mineral soils, but plant establishment is generally better at pH 6.5. If established grassland is maintained at pH 5.5 the clover content is likely to decline compared with a pH of around 6.0. Acidity in the top few centimetres of soil can build up rapidly under high nitrogen usage and is slow to change at depth after lime has been applied to the surface. Thus grassland relying on clover or on high fertiliser nitrogen usage is best limed up to pH 6.0. The lime requirement should not be allowed to rise above 5 t/ha. Regular liming is better than large, infrequent applications.

Soil pH has a large influence on the trace element content of the grass and thereby its nutritional value to the grazing animal. The largest effect is on manganese content. Above pH 6.0, the manganese level in a ryegrass sward may be too low. As shown in Table 15.1, cobalt is also lower at high pH. On the other hand, if high molybdenum uptake is a major problem inducing copper deficiency, a low pH may be beneficial in reducing molybdenum uptake.

Establishment
During the establishment phase, grass mixtures are most sensitive to low pH and deficiency of NPK, especially phosphate. Water-soluble phosphate should be applied if the soil P index is low. Recommendations are given in Table 15.2 for crops sown without a cover crop. Further nitrogen applications may be appropriate during the establishment season depending on sowing date and the extent of intended autumn grazing. For grass/clover crops sown in a cereal cover crop, PK rates not less than those shown in Table 15.2 should be incorporated in

Table 15.1. Effect of pH on nutrient content of a perennial ryegrass ley

	pH 4.7	pH 5.0	pH 5.4	pH 5.8	pH 6.3
	%	%	%	%	%
P	0.31	0.32	0.33	0.34	0.35
K	2.39	2.30	2.35	2.33	2.32
Ca	0.44	0.51	0.58	0.64	0.68
Mg	0.13	0.13	0.15	0.15	0.15
	mg/kg	mg/kg	mg/kg	mg/kg	mg/kg
Mn	282	246	185	135	84
Co	0.17	0.10	0.07	0.06	0.03
Mo	0.22	0.32	0.35	0.40	0.71
Se	0.05	0.04	0.06	0.05	0.05

Source: ADAS.

Table 15.2. Recommendations for grass establishment

	N, P or K index				
	0	1	2	3	Over 3
			kg/ha		
Grass					
Nitrogen (N)	60	60[a]	60[a]	—	—
Grass/legume					
Nitrogen (N)	40	nil	nil	—	—
All crops					
Phosphate (P_2O_5)	120	80	50	30	nil
Potash (K_2O)	120	80	50	nil	nil

[a] Omit nitrogen if sown after end of July.
Source: ADAS.

the cereal seedbed. Up to 50 kg/ha N should be topdressed after the cereal harvest, particularly if the sward is intended for autumn grazing.

Production
If the water supply is non-limiting, grass will respond in terms of dry matter (DM) yield up to very high rates of nitrogen. The rate of fertiliser nitrogen required on a particular field will depend on:
1. Level of DM production determined by climate, soil and sward type (including clover content).

2. System of management.
3. Soil nitrogen supply.

Total DM Nitrogen Response
In the early 1970s, a series of grassland nitrogen experiments was carried out under a range of soil/climatic conditions by the Grassland Research Institute (GRI) and ADAS. From these results established grass growth classes were formed according to water supply for young predominantly ryegrass swards. For each class the potential DM yield and optimum annual fertiliser nitrogen levels were measured. These results are given in Table 15.3. They were produced by cutting experimental plots at grazing frequency. These classes should be reduced by one for land over 300 m. The optimum annual fertiliser nitrogen rates should be reduced by 50 kg/ha or more on fields with a high soil nitrogen supply. This will often apply to fields that have been intensively grazed in previous years. Favoured production areas in the milder, wetter parts of Wales and south-west England are capable of higher production levels than these average levels for England and Wales. The graphs in Figure 15.1 show the difference in response between grazing and silage cutting frequencies derived from these experiments.

Table 15.3. Grass growth classes and potential yields

GRASS GROWTH CLASSES

Soil available water capacity (mm)	April to September rainfall (mm)		
	Less than 300	*300 to 400*	*More than 400*
Less than 100	Poor	Fair	Average
100 to 150	Fair	Average	Good
More than 150	Average	Good	Very good

POTENTIAL YIELD

Grass growth class	*Potential yield (/ha)*	*Optimum annual fertiliser nitrogen (kg/ha)*
Poor	8.4	300
Fair	9.5	330
Average	10.5	370
Good	11.6	410
Very Good	12.7	450

Source: ADAS.

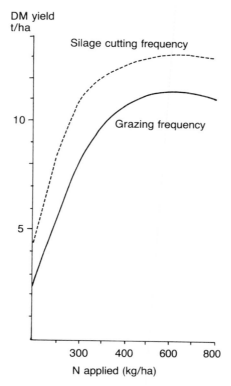

Figure 15.1. Nitrogen response at grazing and silage cutting frequency
Source: GRI/ADAS

The optimum N rate in these studies is defined as the point on the graph at which the slope falls to 10 kg DM/kg N. This is in agreement with recent economic data which value grass at about 3.5 p/kg DM for a range of animal production systems. For the individual farm, this is an average of a wide range of economic values depending on the farming system and the time of year grass is produced. At times, the choice may be between grass or concentrate which will raise the grass value considerably while at other times the grass may not be utilised and so be of no value.

Soil N Supply
The GRI/ADAS experiments also showed that soil N supply and fertiliser N contributions are additive. If the soil N contribution is high, the optimum fertiliser N is reduced. The total N uptake by the crop is

Table 15.4. Effect of soil N supply on optimum annual fertiliser N

	Yield at optimum N t/ha	Soil N supply kg/ha	Optimum fertiliser N kg/ha
4-weekly cutting			
low soil N supply	12.6	16	475
high soil N supply	12.2	102	338
Conservation management			
low soil N supply	15.4	14	445
high soil N supply	13.7	130	220

Source: GRI/ADAS.

closely related to DM yield. All of these experiments were carried out under relatively low soil N supply situations for grassland. Table 15.4 shows the effect of varying soil N supply situations on the optimum annual fertiliser N for sites of similar yields at the optimum. In general the series of experiments shows contributions of about 50 kg/ha N when previous cropping was mainly arable. Old grazed swards might be expected to release two or three times this amount.

Clover

Numerous experiments have been carried out to try to evaluate the clover N contribution of production from mixed swards. The range for ryegrass/white clover has generally been in the range of 100–200 kg/ha N per annum under low fertiliser N use. Persistence of clover depends on several factors, but adequate pH and soil phosphate and potash levels are important. Drought can drastically reduce sward clover content. Newer long petiole varieties of white clover of the Blanca type are generally more persistent at higher fertiliser N levels. Recent experiments have shown a worthwhile clover contribution up to 200 kg/ha fertiliser N. Figure 15.2 from work at Liscombe EHF in west Somerset shows the comparison of ryegrass only and ryegrass/white clover swards at a range of nitrogen levels. Regular defoliation is necessary to maintain the clover contribution to a sward production. It would be unwise to rely on as large a contribution from clover in most commercial farm situations.

Grazed Grass

Most fertiliser experiments on grazed grass are actually performed on grass cut at grazing stage. The grass yield and response to nutrients are measured completely separately from the grazing animal. The reasons

Annual dry
matter yield
t/ha

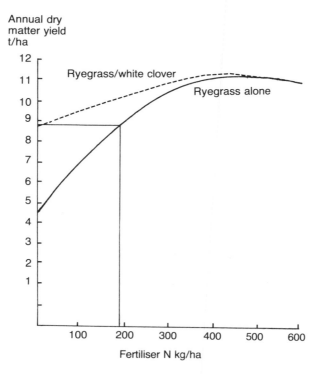

Figure 15.2. Nitrogen response of ryegrass and clover/ryegrass swards
Source: Liscombe EHF, ADAS

for using this experimental approach are readily apparent because it is much quicker, easier and cheaper to carry out. However, it must always be appreciated that fertiliser response measured in this way may not give the same answer as response measured in milk or meat production.

Nitrogen Rate

A series of nitrogen response experiments using cattle live weight gain as the main measure of benefit was carried out by AFRC and ADAS in the 1970s. This series (see Figure 15.3) showed a significant difference in grass dry matter yield between cutting at grazing frequency and actual grazing utilisation by beef cattle. Thus, the optimum nitrogen under actual grazing was 200 kg/ha N lower than that under cutting at grazing frequency. Grazing utilised about 75 per cent of the grass on offer over the season. The lower grazed yields are mainly due to grazing selection, treading damage to the sward and fouling of herbage

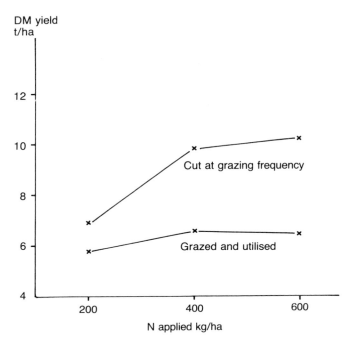

Figure 15.3. Nitrogen response of grazed sward and sward cut at grazing frequency
Source: ADAS

by mud and faeces. Because the design of this series of experiments aggravated some of these factors limiting utilisation, the experiments are currently being repeated with avoidance of these limitations as far as practicably possible. Results to date suggest that utilised grazed grass can respond up to the same nitrogen rates as grass cut at grazing frequency. Live weight gain increases up to these levels have also been shown.

These contrasting results illustrate the practical point that nitrogen response on the farm will only be as good as the actual utilisation of the grass grown. Grazing management needs to be extremely good to benefit from the optimum annual fertiliser nitrogen rates given in Table 15.3. Whether grazing for milk, beef or lamb production, the nitrogen rate must be matched to the grass required. Stocking rate is the major factor in determining nitrogen need.

On intensively stocked dairy farms, fields that are continuously grazed for several years and receiving slurry will provide 100–150 kg/ha N for grass uptake from the soil nitrogen supply. These fields will have an annual optimum nitrogen fertiliser requirement up to 100

Table 15.5. Examples of nitrogen use in intensive grazing systems

	kg/ha N					
Rotational grazing	75	50	50	50	50	50
21 days extending to 28 days in late season	Feb/ March	mid April	early May	end May	end June	end July
Set stocking	75	65	65	50	40	30
28-day application cycle	Feb/ March	mid April	mid May	mid June	mid July	mid August

Source: ADAS.

kg/ha N less than the figures in Table 15.3, even when grazing utilisation is very high. If nitrogen fertiliser use is too high, these fields will show a high soil mineral nitrogen build-up towards the end of the season. This will be very susceptible to winter nitrate leaching.

Nitrogen Timing

Intensive grazing by dairy cows or beef cattle will generally be on either a rotational or a set stocking system. The applications suggested in Table 15.5 are for these two systems giving a total of 325 kg/ha N per year. This assumes an average growth class site under high fertility conditions. It also assumes no major drought limitation to growth during the season, demanding a cut-back in nitrogen use. Application later than mid August is very unlikely to be worth while in intensive grazing systems.

The annual nitrogen recommendations for grass growth classes in Table 15.3 assume average summer rainfall for a particular class. By definition, some years will be considerably dryer than average resulting in reduced grass growth, particularly at the low end of the soil AWC range. If growth is restricted, crop nitrogen uptake is reduced and the optimum fertiliser nitrogen requirement for the season is reduced. In practice, nitrogen fertiliser applications should not continue to be applied to set stocked fields unless the grass is growing. In rotational grazing, it is more obvious when to omit or reduce a planned application.

For less intensive grazing systems, an early application of 50 kg/ha N is likely to be worth while. Mid-season applications may not be necessary, particularly on swards of high clover content. If mid-season nitrogen is not used, the grazing season can usually be extended by applying 50 kg/ha N in August.

Timing of the first and last applications for grazing causes considerable debate. There has been considerable interest in the T-sum approach to timing of first application as used in Holland. Recent work by ADAS and others has shown a wide range of spring timings from which 90 per cent of maximum yield for early-bite grazing can be obtained. There is no benefit to be gained from use of T-sum compared to a calendar date figure for a particular site calculated on the basis of average temperature data. The variation is from late January in west Cornwall, to mid February for much of England, to early March for the north particularly at higher altitude. Figure 15.4 shows the earliest worthwhile application dates for England and Wales. For higher altitudes, later dates are recommended, for example:

Sea level date	150m date	300m date
8 February	+6 days	+17 days
20 February	+10 days	+23 days

The most important consideration is whether early-bite can be grazed without poaching damage. As the soils involved are invariably still at field capacity in most grassland areas, early application is only applicable on the lighter, freely drained fields. Application is better delayed for first grazing on wetter fields.

At the other end of the season, applications later than the end of August are rarely justified, unless extra September production is valued very highly. In most situations autumn mineralisation plus residues of fertiliser nitrogen will be more than adequate to produce maximum growth during September and October. There is also an increased risk of high nitrate levels in the grass which may be harmful to the grazing animal if late nitrogen fertiliser is used.

Phosphate and Potash
Phosphorus and potassium are only needed in modest amounts by all-grazed swards. At soil index of 1–2 for P and 1 for K, these nutrients are applied to maintain soil levels, allowing for the uneven return of faeces and urine. If slurry is returned, fertiliser PK application is not necessary. For maintenance apply 30 kg/ha of P_2O_5 and of K_2O per year. At higher soil indices, application is not necessary. At index 0 for either nutrient apply generous amounts of fertiliser or organic manure to build up the soil level to index 1. Intensively stocked fields rarely have a low soil potash level. To reduce the risk of hypomagnesaemia, potash should not be applied in the spring unless the K index is 0. Herbage analysis of clean grass taken in May can be

Plate C9. Iron deficiency in apples

Plate C10. Boron deficiency in sugar beet

Crown copyright

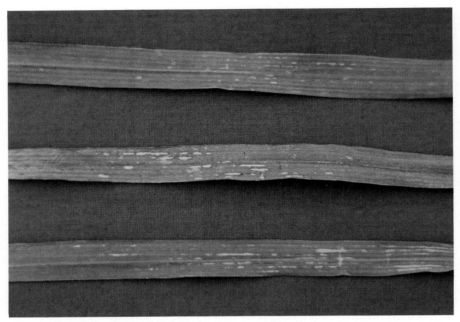

Plate C11. Manganese deficiency in winter wheat

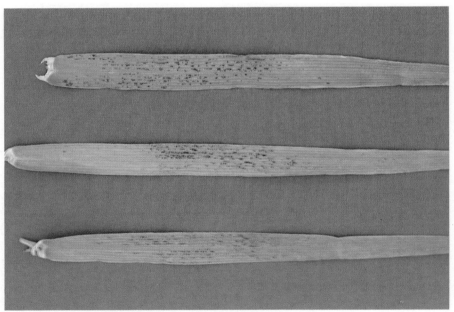

Plate C12. Manganese deficiency in spring barley

Figure 15.4. Mean date for first nitrogen application for early grazing

useful in deciding whether potash use is optimal. The level should be near 2.0 per cent K (DM basis).

Silage

Grass cut for silage removes large amounts of nutrients resulting in high fertiliser recommendations for the crop. The crop requires high rates of nitrogen for optimum response and appropriate rates of phosphate, potash and sulphur to maintain production.

Nitrogen Rate

The nitrogen requirement of a first silage cut depends on yield and soil N supply. Yield depends on the growth stage at which the cut is taken. This in turn determines the digestibility of the final silage. The range of

Table 15.6. Nitrogen recommendations for grass cut for silage

HIGH FERTILITY FIELDS

Long leys and permanent pasture receiving more than 250 kg/ha N, grazed or cut and grazed

Recommended rate per cut – kg/ha N

Cutting strategy	1st	2nd	3rd	4th
4 cuts at 68–70 D silage	120	100	60	60
3 cuts at 64–67 D silage	120	100	80	
2 cuts at 60–63 D silage	120	100		

MODERATE–LOW FERTILITY FIELDS

Short leys in arable rotation and long leys and permanent pasture cut only in previous season or receiving less than 250 kg/ha N

Recommended rate per cut – kg/ha N

Cutting strategy	1st	2nd	3rd	4th
4 cuts at 68–70 D silage	120	100	80	80
3 cuts at 64–67 D silage	150	100	80	
2 cuts at 60–63 D silage	150	120		

Source: ADAS.

optimum nitrogen is surprisingly narrow, usually being between 100 and 150 kg/ha N. Current ADAS recommendations for grass cut for silage are given in Table 15.6. Grass growth class may affect the number or size of cuts which are planned. Summer drought may limit growth and reduce the actual nitrogen applications needed. The production of 60–63 D-value silage is not normally recommended although often achieved in practice.

The application of higher nitrogen rates for first cut will usually give little extra yield while increasing the likelihood of lodging. It will also reduce the soluble carbohydrate level and increase the risk of a poor fermentation. Plate 15.1 shows application to grassland by spinning disc machine.

Nitrogen Timing

For most situations, a single time of application for first cut is all that is justified. This should be applied when growth starts in early/mid March for most of England and Wales. A split application with 40 kg/ha N of the total applied in late February may give a small yield benefit on early sites in favourable growing seasons. It is unlikely to give a yield penalty so that this approach may be adopted if desired. In most situations, it will give no advantage over a single application. If a

Plate 15.1. Spinner application to grassland   *UKF Fertilisers Ltd*

split application is used, the main part of the application should still be applied by mid March in most lowland areas. Nitrogen timing by accumulative temperature or by crop growth stage has not been justified for grass cut for silage.

Nitrogen for second and subsequent cuts should be applied as soon as possible after the previous cut has been removed. This is especially important if a dry spell of weather follows cutting. The soil moisture content under the crop will usually decline once the crop has been removed, and this can delay the speed of action of the fertiliser. The latest application date to achieve an economic response is usually early September.

When an autumn cut of silage is to be taken after summer grazing, no more than 60 kg/ha N is likely to be justified. Response to this application may be low in some years, depending on the fertiliser use earlier in the year. It should not be applied later than early September.

Phosphate and Potash
The amount of phosphate applied at indices 2 and 3 should balance that removed in the grass. A first cut of 25 t/ha of fresh grass removes about 35 kg/ha P_2O_5. A good 3 cut yield of about 50 t/ha of fresh grass removes 70 kg/ha P_2O_5. The comparable dry matter yields are 5 and

10 t/ha DM. Double the rates recommended at index 2 should be applied at index 0 to achieve economic yield response and build up the soil level to at least index 1.

Recent work has shown that phosphate should be applied annually. A yield response to fresh fertiliser, either late autumn or spring applied, is commonly shown at index 2 or 3. At index 0, some phosphate should be early spring applied. At higher indices, the annual requirement may be applied in a single application or with each nitrogen application during the season.

Grass cut for silage removes large amounts of potash. The yields of 25 t/ha and 50 t/ha of fresh grass discussed above will remove about 125 and 250 kg/ha K_2O, respectively. These amounts need to be applied at index 1 to maintain production. At index 0, it may be advantageous to increase the rate by 50 kg/ha K_2O, the extra being applied in the autumn after the last cut. At index 2, particularly on heavy soils, production can generally be maintained at rates of 50 kg/ha K_2O below these figures. The full offtake rates applied at this or higher soil indices will increase potash offtake but have little effect on yield. In most situations, maximum economic response to potash can be achieved at index 1. Increased fertiliser rates result in increased crop removal, but may do little to increase soil level.

Because of the problems associated with using soil analysis to measure the crop response to applied potash, an alternative approach has been developed. Analysis of the grass at cutting can be used to judge whether the potash fertiliser policy is adequate. A level of 2 per cent K in young grass (DM) has been shown to give almost maximum yield. More mature grass requires a lower level.

Where soil analysis is used on fields cut for silage, samples should not be taken within 3 months of removal of a large cut. Soil potash levels are commonly lowered by one index immediately after cutting, but recover over a number of weeks.

Potash is best applied more than once during the season. No more than 90 kg/ha K_2O should be applied before the first cut. The remainder should be applied either after cutting or in the autumn. At indices 0 and 1, part of the application should be applied either in the autumn or before first cut.

Cow slurry contains large amounts of potash (see Chapter 11). The main fields needing potash on an intensive dairy farm are those cut for silage. Whenever practicable, silage fields should receive as much of the slurry as possible, especially if they are never grazed because they are away from the buildings. This will enable much better utilisation of the nutrient content compared to its use on grazed paddocks around the farm. Late autumn generally provides the best opportunity for application to silage ground. The risk of silage contamination generally precludes application between cuts.

Sulphur
Recent experimental work has confirmed that multicut silage in western England and Wales commonly responds to sulphur fertiliser. Responses have been found mainly on sandy, light loamy and chalk soils, especially in south-west England and south Wales, which are the areas known to receive low atmospheric inputs of sulphur (see Chapter 7). ADAS experiments in these areas have shown that analysis of grass at the second or third silage cut stage for N:S ratio is the best indicator of whether a sulphur response is anticipated. The wider the ratio, the greater the likely response.

These experiments have also established that significant yield responses are only likely in second and subsequent cuts. The sulphur supply for the first cut is adequate due to accumulation of soil sulphur from the atmosphere over winter. A large number of recent sites in responsive situations gave the following yield benefits from sulphur fertiliser: first cut 0.04 per cent, second cut 9.7 per cent, third cut 20 per cent and fourth cut 17.5 per cent. Current recommendations are for the use of 15 kg/ha S where the two cuts are taken and 30 kg/ha for three or more cuts on deficient fields. This may be applied before or after the first cut as gypsum or other sulphate-containing fertiliser. There is also interest in soil and foliar applied elemental sulphur.

Cow slurry contains some sulphur but current experiments indicate that mineralisation is sometimes inadequate to achieve the full sulphur response.

Hay
Between 60 and 80 kg/ha N is recommended by ADAS for a hay cut. This low recommendation compared to that for silage assumes a low dry matter yield. For higher-yielding leys or permanent pasture the recommendations for herbage seed production given in Chapter 18 are probably more appropriate as long as the heavier yield can be dried satisfactorily. As for silage the soil maintenance requirements for phosphorus and potassium should be applied. These are about 7 kg/ha P_2O_5 and 21 kg/ha K_2O per tonne DM. A good average yield is 6 t/ha DM for a ryegrass ley and 4 t/ha DM for meadow hay. At index 0 for P or K apply an extra 50 kg/ha of the appropriate nutrient.

Grass for Drying
Swards cut 6–8 times for grass drying with no inputs of organic sources of nitrogen are the most responsive to fertiliser nitrogen. The amount

justified depends on the dry matter yield. In high-yielding situations, a total of 500 kg/ha N is economically justified, applied in increments in spring and immediately after each cut. A generous first application of up to 150 kg/ha N should be applied in the drier parts of England. Phosphorus and potassium use should balance offtake, 7 kg/ha P_2O_5 and 25 kg/ha K_2O per tonne of DM.

Sources of Nitrogen

Neither ammonium nitrate nor urea are perfect nitrogen fertilisers for grass. Recent IAGR work has shown appreciable losses of gaseous N by denitrification following application of ammonium nitrate to warm, wet soils. Recent work on urea has emphasised that no more than 80–100 kg/ha N should be applied in one application to minimise gaseous loss of ammonia. For the same reason, it should not be used for summer applications, when the risk of a warm drying spell of weather following application is high (see Chapter 9).

Injected aqueous ammonia is often used on grassland. It is generally only cost effective if applied at high application rates. Commonly 350 kg/ha N is injected in one early spring application to last the season. Experiments have shown that this may not give as good a balance of grass production as several smaller applications during the season. While the nitrogen is more slowly available than from other materials used at the same rate and timing, a single application frequently fails to give the growth from July onwards that the same rate in split applications would give. It is more suited to a two-cut silage field than for full-season grazing.

In recent years there has been interest in November–December application of aqueous ammonia using a nitrification inhibitor such as nitrapyrin (N-SERVE). While the inhibitor does delay nitrification, variations in temperature and moisture over the next two to three months means that this is less reliable than early spring application without the inhibitor.

HILL GRASS

Economics dictate that any lime or fertiliser applied in a hill situation must be carefully considered to ensure that it is likely to increase animal production. Rather than maintaining a particular pH or phosphate index for optimum sward growth, it is generally more appropriate to apply nutrients at a lower rate but over a larger area to get the best whole farm improvement. In all hill situations, the climatic limitation on production is important. Temperature limits the growing season to as little as four months, but water is rarely limiting.

Most hill land topsoils are inherently low in pH and major nutrients, including available nitrogen. The indigenous sward reflects the soil type. The most productive is *Agrostis/Festuca* which occurs mainly on brown earths. Inputs of lime and NPK will generally give large increases in production. Many acid peats are *Molinia/Nardus* dominant. Much hill land is *Calluna* dominant. Neither of these is very responsive to lime and fertiliser.

Where hill soils are suitable for improvement either by improved management of the indigenous sward or by reseeding, lime and fertilisers are an essential part of the technique. On mineral soils, 5 t/ha of lime is usually enough to raise the surface pH and enable sown seeds to establish. Water-soluble phosphate is also necessary as the P level is usually in the bottom half of index 0. Potash is usually only essential on peats. At establishment, 40–50 kg/ha N is usually worth while. Once the pH and P status has been improved to pH 5.5 in the surface (5.0 for peats) and P index 1, clover will make a significant contribution to production.

To maintain improved upland pasture, regular inputs of lime and fertiliser are needed. Lime is best applied every 3–4 years to ensure that the top 7.5 cm of soil has a satisfactory pH. Regular phosphate applications of 30 kg/ha P_2O_5 per year are also usually essential. This may be applied every 3 or 4 years as long as the amount is appropriate—90 or 120 kg/ha P_2O_5, respectively. Maintenance applications of potash are only justified on index 0 soils. Depending on the amount of clover in the sward and the intensity of grazing, nitrogen application will vary. It is usually limited to 50 kg/ha N in the spring and a similar application in the late summer.

LUCERNE AND RED CLOVER

No nitrogen fertiliser need be applied for establishment or production of these legumes. PK requirements are similar to grass depending on yield and maturity at cutting. Lucerne is sensitive to pH and should only be grown on calcareous soil profiles. It is also sensitive to boron deficiency. If the soil water soluble boron level is below 0.5 mg/l, boron should be applied before sowing. This is only likely on sandy soils. Both lucerne and red clover can fix in excess of 200 kg/ha N per year. Rhizobium inoculation may be necessary when growing these legumes for the first time.

FORAGE MAIZE

By the time this crop is sown in May, most soils will have a high level of mineral nitrogen present, due to spring mineralisation. Nitrogen fertiliser requirement is only 40–60 kg/ha N. Topdressing after emergence is likely to cause crop scorch if prills lodge at the base of the leaf. While 40 kg/ha P_2O_5 is enough for crop response and soil maintenance at P index 1–2, a 12 t/ha DM crop will remove 150 kg/ha K_2O. This may be replaced by fertiliser or, more commonly, the field to be sown to maize receives slurry in the winter or spring. The crop is tolerant of large applications of manure before seedbed preparation.

FODDER BEET AND MANGOLDS

Fodder beet and mangolds respond up to 125 kg/ha N following a cereal. Around 50 kg/ha P_2O_5 and 75 kg/ha K_2O are adequate to balance offtake of modest yields. Under high-yielding situations, up to twice these levels will be needed to balance offtake. Both crops are responsive to sodium. Salt should be applied at 400 kg/ha (150 kg/ha Na) particularly on sandy soils. This should be applied several weeks before drilling to avoid adverse effects on plant population.

SWEDES AND TURNIPS

These crops are traditionally very responsive to phosphate. This is partly due to their being grown on soils in P index 0 much more commonly than other root crops. Swedes in particular do respond very well to phosphate at this soil level and need 150 kg/ha P_2O_5 in the seedbed. At higher indices 75 kg/ha of both P_2O_5 and K_2O are appropriate. Swedes respond up to 75 kg/ha N and turnips to 100 kg/ha N. Swedes are sensitive to boron deficiency.

KALE

Depending on previous cropping, kale will respond up to 125–150 kg/ha N. If grazed, only 50 kg/ha P_2O_5 and K_2O are needed at index 1–2. If the crop is carted off and no slurry is returned, the full offtake should be replaced. For a 60 t/ha crop this will be 70 kg/ha P_2O_5 and 300 kg/ha K_2O.

FORAGE RAPE, STUBBLE TURNIPS AND FODDER RADISH

Apply 100 kg/ha N for early-sown crops following a cereal. If sown later than mid August 75 kg/ha N is adequate. If grazed apply 50 kg/ha P_2O_5 and K_2O. If the crop is carted off and no slurry is returned, at least 100 kg/ha K_2O needs to be replaced.

OTHER CATCH CROPS

Autumn-sown Italian ryegrass, rye or triticale justifies 25–50 kg/ha N in the seedbed depending on previous cropping and sowing date. Earlier-sown crops justify more. Apply an early-spring topdressing of 75 kg/ha N in February for early-bite grazing.

Chapter 16

CEREALS

OVER THE past decade there has been enormous farmer interest in fertilisers for cereals. During this period, yield levels have steadily increased as varieties and disease control have improved. These yield increases would not have been achieved without substantial increases in nitrogen use. These factors have allowed yields of 10 t/ha of winter wheat to be commonly achieved on commercial fields, where the soil is of a high yielding potential. During this decade winter barley has gained in relative importance as a crop in the UK and much experimental work has been devoted to its nitrogen requirement. In this chapter nitrogen use for each of the cereal crops is discussed followed by sections on liming, phosphate, potash and other nutrient needs of cereals.

NITROGEN

Several systems for determining nitrogen recommendations are currently in use in the UK. Most are aimed primarily at predicting the nitrogen topdressing requirement of a winter wheat crop. To provide a basis for understanding the various approaches that are being adopted the next section outlines the principles involved and the main factors influencing fertiliser nitrogen need.

Demand and Supply

Crop uptake = soil nitrogen supply + (fertiliser nitrogen × efficiency of use).

This simple equation is the key to calculating the amount of fertiliser needed by a particular crop. Whatever system of recommendations is adopted, these will be the component parts, although each will not necessarily be quantified. This is usually the case when a classic nitrogen response curve and economic optimum are derived from field experimental work. Research over the last few years has shown that soil and fertiliser nitrogen supplies can be considered as additive. While an efficiency of use factor should also be included in the soil part of the equation, it provides little help in producing practical recommendations as soil nitrogen supply is itself imprecise.

Crop Uptake
This is generally measured as the above-ground uptake of nitrogen by the crop, which reaches a maximum between flowering and harvest. Harvest index, the percentage of the above-ground dry matter production in the grain, is relatively constant at around 50 per cent (dry weight) unless individual grain weight is low due to late drought or disease affecting grain fill. As the concentration of nitrogen in the total dry matter is also reasonably constant total crop uptake is broadly related to yield. A number of studies have shown that at the higher yield levels, a crop uptake of 20–25 kg nitrogen per tonne of grain (85 per cent DM) is likely. Around 75 per cent of this nitrogen is in the grain at harvest.

Both grain and total nitrogen contents will continue to increase with increasing fertiliser nitrogen application above the economic optimum determined by grain yield. There is no precise grain nitrogen content that correlates with either optimum or maximum yield.

The principle remains that high-yielding crops take up more nitrogen than lower-yielding crops. Supplying adequate nitrogen is only one aspect of achieving the desired high yield.

Factors Affecting Soil Nitrogen Supply
Soils contain much total organic nitrogen in the topsoil. Arable topsoils will contain 5,000 kg/ha nitrogen or more. This is constantly undergoing biological change at varying rates for its particular components. Our main interest is in changes that will produce mineral nitrogen for crop uptake (net mineralisation). On long-term arable soils, the nitrogen balance of the previous crop is of primary importance in determining soil nitrogen supply. Some mineral nitrogen will remain in the soil after harvest of the previous crop, but this is usually small unless the crop received a high level of fertiliser and used it inefficiently. The main variation between crops is in the amount of nitrogen left on the field in crop residues. These will be broken down by soil micro-organisms. If the nitrogen content of these residues is above 2 per cent, considerable mineral nitrogen will be released. Residues of oilseed rape and peas may fall in this category whilst cereal straw, which contains less than 1 per cent nitrogen, contributes very little.

Soil type is important because it influences several aspects of the nitrogen cycle. The amount of mineral nitrogen leached over the winter is an important influence on soil nitrogen supply to the next crop. Excess winter rainfall interacts with soil type in determining the extent of leaching. Only when the excess is less than 100 mm, is above-normal mineral nitrogen likely to remain in the profile until the spring, resulting in appreciable difference in soil nitrogen supply. By April the crop will generally contain 50–80 kg/ha nitrogen, most of which will have come from the soil nitrogen supply. This is a quarter to a third of total uptake.

Efficiency of Fertiliser Use
Not all of a nitrogen fertiliser topdressing is taken up by the cereal crop. Commonly between 50 and 60 per cent of the fertiliser is found in the above-ground crop at harvest. Of the rest some will be in the roots, some in the soil biomass and some will be lost by leaching or to the atmosphere.

Variation in efficiency of fertiliser use is a major reason for unexplained variation in economic optima shown in nitrogen response experiments. For example, at 50 per cent efficiency the optimum level of topdressing on a field might be 200 kg/ha nitrogen. If the efficiency were 60 per cent instead, the optimum would be 167 kg/ha nitrogen. Measured differences in the recovery of fertiliser nitrogen using N–15 are commonly greater than this, often without an obvious explanation of whether fertiliser nitrogen was lost or left in the soil unused. The ability of the crop root system to take up the fertiliser, which may be influenced by root disease or dry soil, is likely to be partly responsible for this variation. Some of the differences in recovery of fertiliser nitrogen are likely to be related to soil type and water supply to the crop.

Loss of nitrogen very soon after fertiliser application is a difficult factor to quantify, but is becoming better understood following recent research. Of the commonly used fertiliser materials, urea can lose ammonia direct to the atmosphere and ammonium nitrate can lose part of its nitrate as nitrous oxide and gaseous nitrogen following denitrification.

Basis for Recommendation Systems
It is possible to separate two basic approaches to the prediction of the nitrogen fertiliser requirement of a winter wheat crop. One approach is to carry out numerous experiments under as wide a range of soil nitrogen supply situations as possible, group the sites according to similar characteristics and recommend the average results from this

grouping. The strength of the approach is that there is little argument about the result from one experimental response curve. The answer is provided as an economic optimum amount of fertiliser to apply in that field. However, when individual sites are grouped to provide averages, the recommendation figures in the tables conceal much unaccounted variability.

The main limitation to this approach is coping with all the variables involved and the interactions between them. The result as used in the ADAS system is to rely on a few major factors only, such as soil type, previous cropping and yield level. Use of the N–15 technique to measure both soil nitrogen and fertiliser contributions to crop nitrogen uptake is being used by ADAS to verify and improve the current nitrogen index system, while maintaining the same philosophy in approach.

The other basic approach is to produce a mathematical model of the interrelationship of the factors involved and then test the answers from this model with field experimental data. Work at Rothamsted in conjunction with ADAS Soil Scientists is currently aimed at a much more sophisticated mechanistic model which, using crop, soil and meteorological input data, aims to produce an improved system. This approach will provide a more accurate system by taking account of the numerous processes involved, up to the time of nitrogen application. What happens later in the life of the crop will remain unknown. One practical problem is minimising the amount of input information needed for each field. This means getting the right balance between scientific accuracy and practicality.

This work is adding considerably to our understanding of the factors affecting nitrogen behaviour in the soil. Its main impact on recommendations is likely to be in the more complex previous cropping situations where few empirical field experiments have been carried out. It will also allow us to sort out which field factors are important and which are less so. This causes considerable debate at present, hence the variation in input information between the various systems used in this country and elsewhere.

Accuracy of Prediction
The value of our endeavours in attempting to improve nitrogen recommendations must be tempered by what is biologically possible. Some of the variation in crop use of nitrogen will always be beyond our control. But consistency of yield has improved over the last decade. This is a major step forward, as we need a reliable target yield if we are to improve our nitrogen prediction. An alternative is to plan for a particular nitrogen uptake and accept that the weather and associated

factors may modify final yield. If prediction gets within 20 kg/ha nitrogen of the right answer, that is the best that can reasonably be achieved.

Winter Wheat

ADAS recommendations for the nitrogen fertiliser topdressing of winter wheat are given in Table 16.1. The factors taken into account are yield level, soil type and nitrogen index as described above. At present these recommendations do not attempt to take into account variation in soil nitrogen supply due to rainfall or temperature during the winter, except to suggest a reduction of 25 kg/ha N if the excess winter rainfall is less than 100 mm.

High nitrogen usage can cause problems if no account is taken of its side effects. Traditionally lodging has been of major concern. Current varieties with growth regulator used as appropriate are fairly resistant to lodging so this is of less practical concern but should not be

Table 16.1. Nitrogen topdressing recommendations for winter wheat

Soil type	N index 0	1 kg/ha	2
Yield level up to 7 t/ha			
Sandy soils	175	150	75
Shallow soils over chalk or limestone	175	150	75
Deep silty soils	150	50[a]	Nil
Clays	150	75	Nil
Other mineral soils	150	100[a]	50
Peaty soils	50	Nil	Nil
Organic soils	90	45[a]	Nil
Yield level 7–9 t/ha			
Shallow soils over chalk or limestone	225	200	125[b]
Deep silty soils	200	100[a]	Nil
Clays	200	125	Nil
Other mineral soils except sandy	200	150[a]	100[b]
Yield level above 9 t/ha			
Shallow soils over chalk or limestone	275	250	175[b]
Deep silty soils	250	150	50
Clays	250	175	50
Other mineral soils except sandy	250	200	150[b]

(a) Increase by 25 kg/ha where harvesting of the previous crop has damaged soil structure.
(b) Nitrogen response at N index 2 is very variable. Recommendations of 100 kg/ha N or more should only be used if lodging is considered a low risk.
Source: ADAS.

Plate 16.1. Pneumatic application to cereals   *UKF Fertilisers Ltd*

completely forgotten. Of more current concern is disease. Foliar diseases, especially mildew and *Septoria* species, are often encouraged by the crop microclimate associated with high nitrogen use. This results in the need to apply late season fungicides in conjunction with nitrogen to achieve the best yield. If foliar disease control is inadequate, increasing nitrogen can often produce a marked fall in grain size due to increased disease. By contrast take-all root disease is often less devastating where high nitrogen is used. This is because the poorer root system associated with take-all uses fertiliser less efficiently. Plate 16.1 shows pneumatic application to cereals using tramlines.

Time of Application
Nitrogen applied in the seedbed or during the autumn is not generally worth while, although many autumn cereals receive a low nitrogen compound in the seedbed. In most situations the soil nitrogen supply is more than enough to meet the small needs of the crop during the autumn, whether early sown or late sown. The idea that nitrogen will compensate for late drilling is not tenable. Sometimes a direct drilled crop will show a small response, due mainly to lower mineralisation in undisturbed soil. Incorporation of straw will immobilise soil nitrogen, but recent work has not justified a general need for fertiliser to compensate for this. The soil can usually supply enough for both crop and straw breakdown, except on very sandy soils.

Much has been written about timing of spring topdressing. In most cases experiments struggle to show significant differences in yield for a wide range of timings. Timing has to satisfy three needs:

1. To ensure an adequate supply for crop growth and dry matter production.
2. To stimulate growth at particular stages of crop development.
3. To enable fertiliser to be used as efficiently as possible.

Requirement 1 can be best appreciated by examining Figure 16.1 which shows the cumulative nitrogen uptake curve for a wheat crop. Where soil nitrogen supply is high no fertiliser is needed before early stem extension in April. But if the soil nitrogen supply is low an application of 40 kg/ha N in February/early March is advisable. This will commonly give a worthwhile yield increase as shown in Figure 16.2. Two individual applications are the maximum needed to satisfy this requirement but more may be applied if desired.

Requirement 2 is mainly concerned with ensuring an adequate ear number at harvest. A main application at early stem extension (GS 30) is the best way to ensure as high a tiller survival as possible. If the crop has a low plant population or is backward in growth stage after the

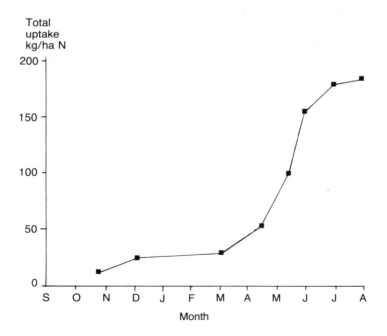

Figure 16.1. Cumulative nitrogen uptake by winter wheat

Plate C13. Manganese deficiency in sugar beet

Plate C14. Manganese deficiency in oilseed rape

Plate C15. Copper deficiency in spring barley   Crown Copyright

Plate C16 (*below*). Molybdenum deficiency in cauliflower
Crown Copyright

winter, early nitrogen will help to encourage more tillering and growth. More than 40 kg/ha N may be worth while, applied as spring growth starts.

Following work at Brooms Barn Experimental Station, there has been considerable recent interest in the precise timing of the main topdressing application. The hypothesis put forward is that the main application should be applied at a precise stage of crop development. This terminal spikelet stage is close to but not necessarily the same as Zadoks GS 30, early stem extension. Recent ADAS experiments have shown that the terminal spikelet stage is not crucial and no better than GS 30 when measured in terms of crop yield response. One practical advantage of the Brooms Barn approach is the ability to model and computer predict the date when the terminal spikelet stage will occur. This enables the nitrogen fertilising of large winter wheat acreages to be planned ahead. GS 30 as currently defined offers a satisfactory guide to optimum timing, but is less easily predicted in advance.

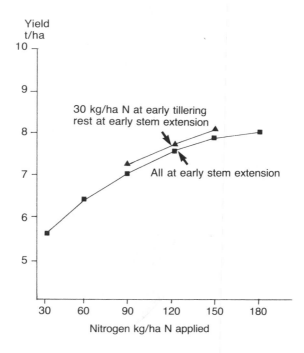

Figure 16.2. Comparison of with and without 30 kg/ha N in late February/early March
Source: ADAS

Requirement 3 demands that the fertiliser is applied to minimise the risks of leaching, denitrification and gaseous loss. It is also important that the surface soil water content is adequate for the crop to take the nitrogen up as fast as it is needed. For ammonium nitrate these criteria demand that fertiliser is not applied to saturated soil. No more than 40 kg/ha N should be applied early unless there is strong justification. One of these situations may be where take-all disease levels are known to be high. The main application should not be applied later than the

Table 16.2. Effect of an extra 40 kg/ha N on protein content of winter wheat

Growth stage	Increase in protein (86 per cent DM basis) (%)
Tillering	Nil
Early stem extension	+0.5
Flag leaf emergence	+0.6
Milky ripe (urea foliar spray)	+0.7

Source: ADAS.

end of April to minimise the risk of slow uptake due to dry topsoil during the high demand period in May. There is a strong case for applying part of the main application earlier in the drier parts of the country if the soil moisture deficit starts to build up in late March. Equally where cereals are grown in high rainfall areas, some of the main application may be applied in May to reduce the risk of denitrification. In summary, most wheat crops will justify 40 kg/ha N early and the rest in April. If logistics demand that the main application is split, summer rainfall is a guide as to whether to apply early or late.

Protein
Milling varieties need a satisfactory grain protein content to be sold at a premium. Additional nitrogen applied at either early stem extension or at flag leaf emergence will produce similar increases in protein content, but the later application is less likely to give a yield advantage. As shown in Table 16.2, nitrogen applied at early tillering will give a much smaller increase in protein content compared to a similar quantity of fertiliser at early stem extension. The later nitrogen is much more likely to be translocated to the grain, rather than remaining in the straw.

Application of nitrogen rates higher than those recommended for yield will often give a further increase in protein content, particularly for high-yielding crops. This may enable the crop to be sold for a higher price. Some milling wheat is sold for a premium if it is above a particular threshold protein content rather than on a sliding scale. There is no guarantee that a particular nitrogen rate will ensure that the grain is above a particular threshold.

Foliar sprays of urea during grain filling can be used to increase grain protein content, but there is a risk of scorch. This may or may not affect grain yield. The protein benefit is only a little higher than that from the same amount of fertiliser nitrogen applied earlier.

Spring Wheat
There has been little recent work on this crop. In general, recommendations should follow those for winter wheat yielding up to 7 t/ha. This includes spring wheat varieties sown in either late autumn or spring.

Nitrogen should be applied as a split application unless early nitrogen is not justified due to high soil levels. The first application of 40 kg/ha N should be applied in late February/early March for late autumn-sown crops, topdressed in March at the 3 leaf stage for January/February drilled crops or applied in the seedbed of later drilled crops. The rest of the nitrogen requirement should be topdressed at early stem extension (GS 30) and/or by the end of April.

Little work has been carried out on the benefits, if any, of more or later nitrogen applications on grain protein content. If protein levels are not satisfactory, it is reasonable to assume that 40 kg/ha N at flag leaf emergence would achieve a higher protein content and be worth considering.

Durum Wheat
Recommendations for winter wheat may be too high if lodging is likely to be a problem with a particular durum variety. Otherwise these levels or slightly above are appropriate.

Winter Barley
Winter barley is very similar to winter wheat in its nitrogen requirements. The main differences which modify aspects of both rate and timing are:

- lower yield potential,
- susceptibility to lodging,
- more vigorous autumn and early spring growth,
- malting quality of some varieties.

Lower yield potential will limit the overall nitrogen demand of the crop. Lodging susceptibility may limit nitrogen response compared to winter wheat. The ability of the crop to grow vigorously during autumn and early spring may influence nitrogen timing while the low grain nitrogen needed for malting will influence both rate and timing of nitrogen fertilisers. ADAS topdressing recommendations for winter barley are given in Table 16.3.

Table 16.3. Nitrogen topdressing recommendations for winter barley

	N index		
	0	*1*	*2*
Soil type	kg/ha		
Sandy soils, shallow soils over chalk or limestone	160	125	75
Other mineral soils	160	100	40
Peaty soils	50	Nil	Nil
Organic soils	90	45	Nil

Source: ADAS.

During recent years, both rate and timing have been examined in detail by ADAS. All sites followed at least one previous cereal. For each site the economic optimum nitrogen topdressing level has been assessed. While the average is near 160 kg/ha N, the current ADAS recommendation, several sites responded to 200 kg/ha N but few above this level. Many of the low optima occurred in 1981, when yields were generally poor and net blotch was a particular problem.

While it is not possible to provide precise guidance on which crops justify the 200 kg/ha N application, some factors are common. In general these are long-term arable fields, often in continuous cereal and on medium textured soils. Frequently a high yield was achieved, but lodging rarely occurred. Growth regulator was used on a number of sites. The key factors are a site of good yield potential but very low soil N supply and lodging risk. A level of 200 kg/ha N should only be used on varieties with reasonable standing power.

Ear size varies considerably in winter barley varieties. This is reflected in the very different ear numbers needed for high yields of different varieties. For Maris Otter, Igri and Sonja, ear number tended to increase up to 180 kg/ha N. (Figure 16.3.)

Yield level has not been possible to separate out as a variable in the winter barley recommendations. Examination of an early series of ADAS experiments showed that season, variety, soil type and yield

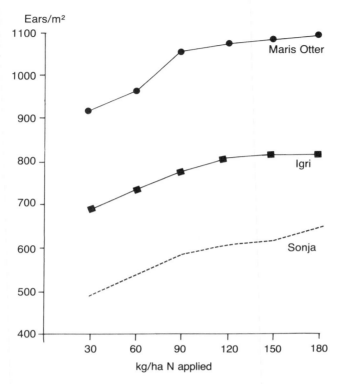

Figure 16.3. Effect of nitrogen level on ear number of three winter barley varieties
Source: ADAS

level were not able to be distinguished as factors affecting the optimum rate of topdressing. Nitrogen index gave a difference in average optimum of 40 kg/ha N.

Lodging of winter barley is common and the risk is increased with higher nitrogen rates. In an earlier series of experiments, all but one site lodged in 1977 and around half the sites in each year 1978–80 showed some lodging at least at the highest rate of nitrogen tested. The average optima showed a clear separation of lodged and unlodged sites.

              kg/ha N
Not lodged: 176 (38)
Lodged: 134 (38)

    Number of sites in parentheses.

Examination of the site data showed no overall increase in lodging due to yield level, variety, soil type or ear number. Even N index was equally distributed between lodged and non-lodged sites. The inference is that lodging is due mainly to weather factors, particularly wind and rainfall intensity.

Nitrogen and Growth Regulator
A number of experiments in the last few years have looked at the effect of growth regulator use for lodging control on nitrogen response. While the use of growth regulator will increase the likelihood of an optimum nitrogen rate of 200 kg/ha N rather than 160 kg/ha N, this extra yield response is unlikely to pay for both the growth regulator and the extra nitrogen. The growth regulator must give some yield improvement in its own right before the combination becomes cost effective. This leaves the use of extra nitrogen and/or growth regulator as two separate decisions depending on the particular field and crop situation. Usually the decision on growth regulator use for lodging control should be taken first. The appropriate nitrogen rate can then be decided.

Timing
In recent ADAS experiments, application of part of the nitrogen in the seedbed and in February was compared with all applied at early stem extension (late March/early April). In some sites seedbed N contributed to total N uptake but rarely had a unique effect on yield. As long as adequate spring N was applied, seedbed N was not needed and is not recommended because crop use is less reliable than spring-applied nitrogen. Application of a two-way split topdressing with 40 kg/ha N applied in February was generally beneficial when followed by the main early stem extension (GS 30) application at the normal timing in late March or April. Recent Long Ashton work has shown that March applications can increase foliar disease levels even in the presence of a fungicide programme. Delaying the main application until April can be advantageous in some situations where the disease pressure is high.

Recommendations for timing of applications are as follows:

1. No seedbed or autumn nitrogen.
2. Apply about 40 kg/ha N during tillering but not before mid February in most situations. This applies to all feed crops, whether forward or backward in growth. The bigger yield advantage will occur on the backward crops.

3. For feed varieties, apply the rest of the topdressing at around early
   stem extension (GS 30)—usually between late March and mid
   April. Topdressing should be completed by late April.
4. For malting varieties complete topdressing by mid March. In many
   situations a single application will be most appropriate.

Malting
Maris Otter and other varieties grown specifically for malting generally
need a relatively low grain nitrogen content to attract a financial
premium. If this is the desired market, nitrogen application may need
to be reduced, resulting in both lower grain N and lower yield in most
circumstances. Due to site factors, low grain N contents are only likely
to be reliably achieved on sandy, light loamy and shallow chalk soils
with low organic N residues from previous crops. Even in these
circumstances, rainfall distribution in a particular season may cause
late nitrogen uptake, resulting in too high a grain N content. For
malting the aim is usually 1.6–1.8 per cent N (DM basis) depending on
the particular market. Mean grain N contents for sixteen Maris Otter
sites 1978–80 were as shown in Figure 16.4. As the data show, the

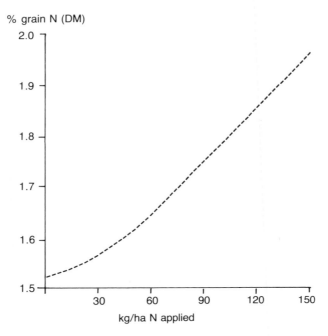

Figure 16.4. Effect of nitrogen on Maris Otter grain nitrogen content
Source: ADAS

increase in grain N tends to be linear above 100 kg/ha N applied. For the thirteen sites where early N was applied by mid March, timing gave the following mean result for all N rates:

All mid March	All mid April	50:50 split
1.74	1.83	1.75

These figures confirm the advantage of applying at least part of the total N early to keep grain N low. For these thirteen sites the rate of response to N over the range 90–150 kg/ha N was 9 kg/ha N. Therefore reducing the N rate from 150 to 100 kg/ha N in the hope of producing premium quality will reduce yield by an average of 0.45 t/ha.

Spring Barley

Recent work has shown that spring barley nitrogen recommendations also justify being increased as has happened with winter cereals. Higher-yielding varieties and much improved lodging resistance contribute to higher optimum nitrogen requirements. ADAS recommendations for the crop are given in Table 16.4. Spring cereals generally make poorer use of soil nitrogen compared to winter cereals. Some of the nitrogen that the winter cereal takes up during the autumn and spring is leached before the spring cereal is drilled. Therefore the difference between nitrogen recommendations for spring barley at indices 0 and 1 is generally less than for winter cereals.

Timing

On sandy and shallow soils, the risk of nitrogen fertiliser being leached below rooting depth is high for spring-sown crops. Nitrogen should generally be applied as a split application unless early nitrogen is not justified due to high soil levels. The first application of 40 kg/ha N

Table 16.4. Nitrogen recommendations for spring barley

	N index		
	0	*1*	*2*
Soil type		*kg/ha*	
Sandy soils	125	100	50
Shallow soils over chalk or limestone	150	125	50
Other mineral soils	150	100	40
Peaty soils	40	Nil	Nil
Organic soils	70	35	Nil

Source: ADAS.

should be topdressed in March at the 3 leaf stage for January/February drilled crops or applied in the seedbed for later drilled crops. The remainder should be topdressed at the early stem extension (GS 30) stage and/or by the end of April. On soils where the leaching risk is low, crops drilled from March onwards can receive a single application in the seedbed.

Malting
To achieve a satisfactory grain nitrogen content for malting, nitrogen applications should be completed by early tillering or mid March, whichever is the later. Applications slightly below those recommended for yield will increase the likelihood of achieving the desired grain nitrogen content. No more than 120 kg/ha N is recommended on N index 0 fields.

Oats
The ADAS recommendations are given in Table 16.5. These apply to both winter- and spring-sown crops. Recent work has confirmed these figures and re-emphasised lodging as the major factor limiting response to nitrogen. Delaying some nitrogen until GS 35–37 did not reduce lodging in winter oats. Due to the high lodging risk with this crop, it is recommended that early nitrogen is only used on low lodging risk sites. For the majority of situations, nitrogen should be given as a single application at early stem extension (GS 30).

Rye
ADAS recommendations are the same as for oats (Table 16.5). The lodging risk is considerable and, even with growth regulator, the crop is usually only grown on very sandy soils.

Table 16.5. Nitrogen recommendations for oats and rye

| | N index | | |
| | 0 | 1 | 2 |
Soil type	kg/ha		
Sandy soils, shallow soils over chalk or limestone	125	100	50
Other mineral soils	100	60	30
Peaty soil	40	Nil	Nil
Organic soils	70	35	Nil

Source: ADAS.

Table 16.6. Phosphate and potash recommendations for cereals

	P or K index				
	$0^{(a)}$	1	2	3	Over 3
			kg/ha		
Straw ploughed in or burnt					
Yield level 5.0 t/ha					
Phosphate (P_2O_5)	90	40	40M	40M	Nil
Potash (K_2O)	80	30	30M[b]	Nil	Nil
Yield level 7.5 t/ha					
Phosphate (P_2O_5)	110	60	60M	60M	Nil
Potash (K_2O)	95	45	45M[b]	Nil	Nil
Yield level 10.0 t/ha					
Phosphate (P_2O_5)	130	80	80M	80M	Nil
Potash (K_2O)	110	60	60M[b]	Nil	Nil
Straw removed					
Yield level 5.0 t/ha					
Phosphate (P_2O_5)	90	40	40M	40M	Nil
Potash (K_2O)	110	60	60M[b]	Nil	Nil
Yield level 7.5 t/ha					
Phosphate (P_2O_5)	110	60	60M	60M	Nil
Potash (K_2O)	140	90	90M[c]	Nil	Nil
Yield level 10.0 t/ha					
Phosphate (P_2O_5)	130	80	80M	80M	Nil
Potash (K_2O)	170	120	120M[c]	Nil	Nil

(a) At index 0 large amounts of phosphate and potash are recommended to raise the soil index over a number of years.
(b) Not needed on most clay soils.
(c) A lesser amount may be used on most clay soils.
M This indicates a maintenance dressing intended to prevent depletion of soil reserves rather than to give a yield response.
Source: ADAS.

Triticale
ADAS nitrogen recommendations are the same as those given for oats and rye in Table 16.5. Most triticale is grown on sandy or shallow soils following a previous cereal. Recent experiments have shown that lodging commonly limits yield above the 125 kg/ha N recommended level.

Table 16.7. Annual versus triennial phosphate and potash for a sugar beet and cereal rotation

		Annual phosphate	Triennial phosphate	Annual potash	Triennial potash
		t/ha grain (85 per cent DM) or sugar			
1973	Sugar beet	9.54	9.45	9.49	9.50
1974	Winter wheat	4.80	4.86	4.74	4.85
1975	Spring barley	4.22	4.29	4.15	4.30
1976	Sugar beet	5.74	5.75	5.76	5.85
1977	Spring barley	5.86	5.85	5.81	5.77
1978	Spring barley	5.64	5.65	5.58	5.53
1979	Sugar beet	7.31	7.64	7.73	7.69
1980	Winter wheat	7.45	7.48	7.47	7.57
1981	Winter wheat	8.86	8.90	8.52	8.66
Average cereal yield		6.14	6.17	6.05	6.11

Source: Norfolk Agricultural Station.

LIMING

While liming to maintain pH 6.5 is the general arable recommendation, there are circumstances where a lower field pH is acceptable. Wheat and oats will grow satisfactorily at pH 6.0, while rye at 5.5 will give satisfactory results. Topdressing with lime is only justified if it is discovered that the pH after drilling is 0.5 pH unit below these values. Barley prefers a pH of 6.5 or above and there is some evidence that pH 7.0 is optimal for this crop.

PHOSPHATE AND POTASH

Cereals will give a yield response to applied phosphate or potash at index 0 and sometimes index 1 for the appropriate nutrient. At these low nutrient levels combine drilling, particularly of phosphate at index 0, will give a larger yield response than broadcasting almost regardless of the rate of application. However the difference in yield is still small relative to delayed drilling date especially of spring cereals. Combine drilling should only be used when no delay in farm drilling date will result.

At indices 1–3 for phosphorus and 1–2 for potassium maintenance levels of application depending on removal in grain and/or straw are appropriate (see Table 13.4). Depending on yield level and assuming straw yield is 70 per cent of grain yield, the ADAS recommendations are given in Table 16.6. No phosphate or potash is recommended at high soil indices.

Maintenance applications are not crucial to yield at index 2 or above and can be applied annually or every other year if more convenient, except for potash on sands. This should be applied annually. Table 16.7 presents data on annual versus triennial application to cereals, showing that each is acceptable on sandy boulder clay once a satisfactory soil index has been achieved.

MAGNESIUM

Although visual symptoms of magnesium deficiency are seen on cereals, poor root activity and low nitrogen uptake are generally responsible. Symptoms are particularly common in spring barley grown on chalk soils. Yield responses to magnesium fertilisers are rare. Cereals are only likely to justify a magnesium fertiliser application on sands below 15 mg/l Mg (low index 0). In this situation, 50 kg/ha Mg should be applied every three to four years.

SULPHUR

Recent experiments on sulphur fertiliser response of milling wheat have shown very small, usually non-statistically significant yield responses. This includes sites in areas of known low atmospheric sulphur inputs where grass cut for silage shows substantial yield responses. Doubt remains over the likelihood of breadmaking-quality benefits. If found, these are also likely to be small judged on current evidence.

TRACE ELEMENTS

Only manganese and copper cause problems. Manganese deficiency is common and should be treated by foliar spraying. In very susceptible fields, autumn application to early-sown winter cereals may be necessary. Otherwise, crops should be treated in the spring as soon as symptoms are seen or when leaf cover is sufficient on known deficient fields. On alkaline organic soils complete kill of plants can occur if treatment is not applied. In these severe situations, two or three applications may be necessary.

Copper deficiency gets less and less common as most deficient fields have been regularly treated for a number of years. Soil analysis or previous experience is needed to decide on treatment as application is best carried out during tillering. Symptoms often do not show until flag leaf emergence. Foliar application is usually more appropriate than soil treatment. Late foliar application to cereals is less beneficial than when applied during tillering and much more likely to cause foliar scorch.

Chapter 17

SUGAR BEET AND POTATOES

WHILE IT is convenient to put the two major root crops grown in the UK in the same chapter, their fertiliser requirements are very different. Each has traditionally received high fertiliser applications, which have contributed appreciably to the building up of soil nutrient levels in arable areas.

SUGAR BEET

Sugar beet responds to modest applications of nitrogen and sodium, but only responds to other major nutrients if their soil index is low.

Liming
Sugar beet is very sensitive to acidity and poor patches of growth due to low pH are commonly seen on sandy soils (see Plate C1). On mineral soils a pH of 6.5 is necessary. On organic and peaty soils 5.8 is satisfactory in the topsoil. Acid peat subsoils need to be limed to 5.0 to a soil depth of 40 cm before beet can be grown successfully. This means that the crop should not be grown on newly limed, acid peat unless some lime has been ploughed down or incorporated below plough depth.

Nitrogen
Many experiments over several decades have confirmed beyond all reasonable doubt that only modest rates of nitrogen are needed for maximum sugar yield. As shown in Figure 17.1 the yield of tops increases up to quite high nitrogen rates, but root yield is less responsive and sugar yield declines if nitrogen usage is too high. This is because the sugar content of the roots falls as nitrogen application is increased.

In the normal rotational position following a cereal crop, 125 kg/ha

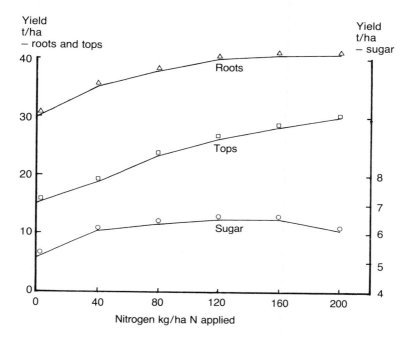

Figure 17.1. Effect of nitrogen on yield of sugar beet roots, tops and sugar
Source: Brooms Barn Experimental Station

nitrogen should be applied on sandy and shallow soils over rock. Medium-textured mineral soils need no more than 100 kg/ha nitrogen. Recent work has confirmed that 50 kg/ha nitrogen is sufficient on Fen peats. These recommended rates do not need to be increased for high yields. Many high yielding experiments have shown an optimum nitrogen level no higher than 125 kg/ha nitrogen. Where the crop is early harvested, slightly lower rates of nitrogen will give maximum sugar yield. If the tops are being utilised for animal feed, the nitrogen rate should not be raised above 150 kg/ha nitrogen or the sugar yield will be reduced.

Recent work has confirmed that the nitrogen requirement is not increased by irrigation. The results in Figure 17.2 from Gleadthorpe EHF show that yield can be increased by irrigation in some years but 125 kg/ha nitrogen is still adequate.

As nitrogen rate increases, the amino–N level in the roots is also increased which results in poorer extraction of the sugar during processing. In some European countries, high amino–N contents

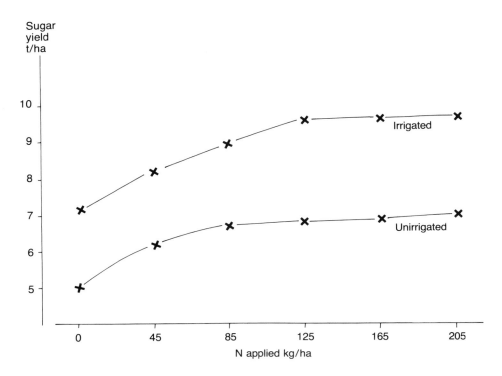

Figure 17.2. Nitrogen response of sugar beet with and without irrigation
Source: Gleadthorpe EHF, ADAS

attract a financial penalty. Work in this country has shown that use of poultry manure for the crop results in very high root amino–N levels.

Timing of Nitrogen
Timing of application is important because:

- nitrogen in the seedbed can reduce plant establishment,
- early nitrogen is more at risk from leaching.

On mineral soils, particularly sandy textures, soluble fertiliser in the seedbed will reduce plant establishment. An adequate plant population is essential to maximum yield, so any risk due to high fertiliser rates in the seedbed should be avoided. The results from Norfolk Agricultural Station shown in Table 17.1 are typical. Leaching would not be a high risk on this medium-textured soil.

The results in Table 17.2, from work on irrigated beet on loamy sand

**Table 17.1. Effect of various methods and timings of
125 kg/ha N for sugar beet on plant populations (1979–81)**

	Population 1,000/ha
No nitrogen	72
125 kg/ha N broadcast	
2/3 weeks predrilling	75
immediately post drilling	75
2 weeks after drilling	68
4 weeks after drilling	67
6 weeks after drilling	75
40 kg/ha N broadcast at drilling and 85 kg/ha N	
2 weeks after drilling	72
4 weeks after drilling	72
6 weeks after drilling	76
125 kg/ha N sideband at drilling	74

Source: Norfolk Agricultural Station.

**Table 17.2. Effect of rate and timing of nitrogen on sugar yield of irrigated sugar beet
on sandy soil (1979–81)**

Timing	kg/ha N		
	125	165	205
	t/ha of sugar		
Seedbed	9.21	9.94	9.46
Post drilling overall	9.84	9.75	9.59
Post drilling deflected away from rows	9.53	10.08	9.43
40 kg/ha in seedbed remainder post emergence	9.80	9.88	9.91

Source: Gleadthorpe EHF, ADAS.

at Gleadthorpe EHF where the leaching risk is much greater, show that application of 40 kg/ha nitrogen in the seedbed and the remainder after full emergence has produced the most reliable result. The wetter the April/May period, the more likely the benefit from delaying the application until full emergence, but not later than the end of May. By this time the crop root system will be deep enough to use nitrogen that has moved down the profile. If all the nitrogen is applied early, a May topdressing of 50 kg/ha N extra nitrogen may be necessary in a wet spring.

On the medium and heavy soils with a much lower leaching risk, use of deflector plates to place the fertiliser between the rows at or soon after drilling has proved very effective in avoiding adverse effects on plant establishment. Seedbed application is satisfactory on peats, as the risk of both reducing plant population and leaching is small.

Phosphorus

Sugar beet only respond appreciably to phosphate fertiliser when the soil P index is 0 or 1. At an index of 2 or above the rates recommended are to maintain soil reserves. Table 17.3 shows the increase in yield due

Table 17.3. Effect of phosphate fertiliser (126 kg/ha P_2O_5) on sugar yield at different soil P indices

ADAS P index	Number of experiments	Mean response (t/ha of sugar)
0	4	+ 1.1
1	9	+ 0.3
2	19	+ 0.2
3	26	+ 0.1
4 +	12	− 0.1

Source: Brooms Barn Experimental Station.

to phosphate fertiliser from experiments carried out by Brooms Barn Experimental Station. The main response occurs at index 0. The recommendation figures allow for crop response at the low indices and maintenance of soil levels at indices 2 and 3. The 50 kg/ha P_2O_5 level is enough to balance offtake of high yielding crops. Up to 70 kg/ha P_2O_5 will be needed if tops are also being removed from the field. Experiments have shown that autumn application of phosphate at P index 2 or above is completely satisfactory.

Potassium and Sodium

For maximum yield, sugar beet need an adequate supply of both potassium and sodium. While some substitution of one cation by the other can occur, both are essential for yield. Sodium is more important on low potash, droughty soils. This covers most sandy soils in sugar beet growing areas. Sodium is needed in smaller amounts on moisture

Table 17.4. Potash response of sugar beet with and without sodium
(20 experiments 1970–74)

	kg/ha K_2O				
	0	*42*	*83*	*167*	*333*
			t/ha of sugar		
Nil Na	5.47	5.67	5.63	5.79	5.99
150 kg/ha Na	5.98	5.88	6.10	6.18	6.17
% yield increase due to Na	9.30	3.70	8.30	6.70	3.00

Source: Brooms Barn Experimental Station.

retentive, relatively high potash status silts and peats. Table 17.4 illustrates the relationship between responses to the two nutrients.

Potash response is high at indices 0 and 1. At index 2 and often index 3, a worthwhile response to potash is given. For this reason a high rate of 200 kg/ha K_2O is recommended at index 0, with 100 kg/ha K_2O at index 1. At indices 2 and 3, 75 kg/ha K_2O is recommended to ensure any yield benefit is produced. At index 2 for high yielding crops, 100 kg/ha K_2O is needed to balance root offtake and maintain soil levels. If tops are also removed, the total potash removal will be 250 kg/ha K_2O. This should be applied at indices 0–2 or the soil level will fall. All these potash recommendations assume that appropriate sodium is also applied.

Potash at index 1 and above may be autumn applied. If potash is spring applied, it should be well worked into the seedbed. On sandy soils where the risk to plant establishment is high, application is better carried out in January or February if possible, so that there is enough rain between application and drilling to leach the chloride part of the fertiliser below the level of the young plant roots. There is a risk of some potassium loss by leaching if all the potash is ploughed down in the autumn on very light soils. Both sodium and potash are best applied after ploughing in these circumstances, preferably in January or early February.

Recent work on sodium response carried out at Brooms Barn on different soil types has emphasised the importance of sodium on sands. Most of these soils had available sodium levels below 20 mg/l Na. By contrast no responses were shown at soil levels above 40 mg/l Na. Most organic soils were in this category. Most loams and silts were in either the 0–20 or the 20–40 category and small responses were shown in

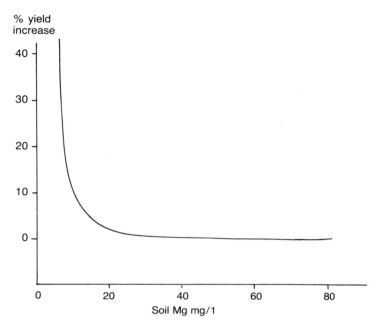

Figure 17.3. Magnesium response of sugar beet
Source: Brooms Barn Experimental Station

both. Fen silts showed few responses, probably due to appreciable subsoil sodium levels in many cases.

On sandy soils 400 kg/ha salt (150 kg/ha Na) is recommended. On loamy soils, half this rate will achieve maximum yield response. Although occasional sites will respond, there is no justification for general use of sodium on peats or Fen silts. If sodium is not used on responsive soils, an extra 100 kg/ha K_2O will partly compensate for the omission. Salt is very damaging to young beet seedlings and along with other soluble fertilisers should not be applied just before drilling. Autumn or winter application is recommended, depending on leaching risk.

Soil analysis for sodium is not generally used as levels of available sodium are very strongly related to soil texture. Sodium levels will not build up in a soil, as it is less strongly held than calcium or potassium. Sodium not taken up by a beet crop will be leached over the next two winters so the soil level is likely to be back to its initial level by the time beet are grown again.

Magnesium

Sugar beet is the only UK crop on which a large number of magnesium response experiments have been carried out. The work was carried out by Brooms Barn Experimental Station during the 1960s mainly on sandy soils in eastern England. Since that time, there has been a marked increase in the use of magnesium fertilisers and it would be difficult to find as many low soil magnesium sites today.

The results produced the relationship shown in Figure 17.3. The large responses occurred around 15 mg/l Mg or less. Very few responses occurred in soil index 1. Following this work 100 kg/ha Mg is recommended at index 0 and 50 kg/ha Mg at index 1. It is debatable whether the latter application is worthwhile above 30 mg/l Mg.

Manganese

Manganese deficiency (speckled yellows) is common in sugar beet crops grown on organic soils and also occurs on high pH sands. The symptoms may be aggravated by post-emergence herbicide damage. The problem is particularly common on seedlings at the two–three leaf stage, before the leaf area is high enough to allow appreciable uptake from foliar spraying.

Work at Brooms Barn has developed the incorporation of manganese oxide in the seed pellet. This has proved very successful in preventing early deficiency. Often the crop will still need treatment later but, by the end of June, the leaf area is much greater and foliar spraying is more successful. Manganese coated seed is now available as an option for use by growers on susceptible soils.

Boron

Boron deficiency still occurs from time to time on sandy soils usually when the pH is above 7.0. Prevention is the only satisfactory policy. Boron is widely available in autumn-applied blended fertilisers and in NPK compounds. Alternatively, boron can be soil applied in the spring. Boron is susceptible to leaching so spring application is ideal. Autumn application is generally satisfactory where an insurance application is required. Foliar spraying is a last resort. The problem usually shows during a period of drought and is rare in fully irrigated crops.

SUGAR BEET SEED

The sugar beet seed crop is usually drilled with a cereal cover crop. For the first year apply a moderate nitrogen rate appropriate to the cereal and rates of phosphate, potash and magnesium as for the sugar beet crop. To avoid excessive nutrients in the seedbed, it is better to apply all except the nitrogen in the previous autumn and topdress at least part of the nitrogen after emergence. Sodium should be applied and this may be topdressed in the autumn after harvesting the cover cereal. On soils of index 2 or above, the phosphate and potash may also be applied at this time if preferred.

As soon as the cereal has been harvested and the straw carted off, topdress the seed crop with about 75 kg/ha N. This should be followed by 200–250 kg/ha N in March/early April of the seed harvest year. Seed yield is very responsive to nitrogen, but no recent experiments have been carried out in this country.

POTATOES

Experiments show that potatoes respond to high applications of phosphate and potash. These responses are economic at a soil index of 3 or above. At these high levels virtually no other crops show a yield response.

Liming
Potatoes will give maximum yields at lower soil pH levels than most other arable crops. Liming prior to growing potatoes should be avoided as it will often increase the incidence of common scab. As long as the soil pH is 5.5 or above on mineral soils and 5.0 on peats, liming should be carried out after lifting the potatoes in readiness for the following crop. If the soil is naturally calcareous and has a high pH, potato yield will not be reduced in most circumstances.

Second Early and Maincrop
The main principles of NPK use are considered in this section. The requirements of these crops when grown for processing are considered later.

Nitrogen
Nitrogen response has been examined over many years. The ADAS recommendations given in Table 17.5 show how soil type and previous cropping (N index) are taken into account. Very few experiments, even on sandy soils, show a yield response above 250 kg/ha N. Water is the primary limitation to potato yield. Where adequate water is given and yields of 60 t/ha or more achieved, extra nitrogen is not required. It would seem that the higher yielding crop uses the nitrogen available

Table 17.5. Fertiliser recommendations for maincrop potatoes

Soil type		N, P or K index				
		0	1	2	3	Over 3
		kg/ha				
All mineral soils	Nitrogen (N)	220	160	100	—	—
	Phosphate (P_2O_5)	350	300	250	200	100
	Potash (K_2O)	350	300	250	150	100
Peaty soils	Nitrogen (N)	130	90	50	—	—
	Phosphate (P_2O_5)	350	300	250	200	200
	Potash (K_2O)	350	300	250	150	100
All other organic, moss and warp soils	Nitrogen (N)	180	130	80	—	—
	Phosphate (P_2O_5)	350	300	250	200	200
	Potash (K_2O)	350	300	250	150	100

Source: ADAS.

more efficiently. The main effect of nitrogen is to increase tuber size.

Under most circumstances all the nitrogen should be applied at or before planting. While this will delay tuber initiation by a few days compared to very little nitrogen at planting, this is unlikely to affect yield adversely by the end of the season. The main circumstance in which seedbed nitrogen may be unsatisfactory is where the fertiliser is leached before the crop has developed its root system. On sandy soils this is an appreciable risk during March and April.

Work at Gleadthorpe EHF on irrigated crops has shown that a split application is more appropriate under these conditions (Table 17.6). By applying half the nitrogen at planting and half at tuber initiation in June, a yield advantage was shown. For irrigated crops on sandy soils this split application is recommended. If very low soil moisture deficits are being maintained for scab control during late May–early June, the topdressing should not be applied until tuber initiation and the change in irrigation practice to a higher soil moisture deficit limit. The Gleadthorpe results show similar yields from a 50:50 split and from a 50:25:25 split. In the latter case 125 kg/ha N was applied in the seedbed with the remaining 125 kg/ha N split between tuber initiation and four weeks later in July. As long as a full irrigation regime is maintained so that the very late nitrogen is not left on the dry soil surface this policy might be considered. It is most appropriate on very light soils if the water application is likely to be generous but uneven and should only be adopted for a crop grown to maturity.

**Table 17.6. Effect of rate and timing of nitrogen on ware yield of
irrigated maincrop potatoes on sandy soil (1977–79)**

N applied kg/ha	All on seedbed t/ha	½ seedbed ½ tuber initiation t/ha	½ seedbed ¼ tuber initiation ¼ 4 weeks later t/ha
125	57.0	59.9	61.5
250	58.8	65.4	63.8
375	58.8	62.7	61.4

Source: Gleadthorpe EHF, ADAS.

Phosphorus

While potatoes are very responsive to phosphate fertiliser, the rate of
response declines as the soil P index increases. At P index 1–2 the
response is commonly 20–30 kg tubers/kg P_2O_5. At index 3, 10 kg
tubers/kg P_2O_5 is common while at 4–5 the rate of response is small at
0–5 kg tubers/kg P_2O_5 applied. Figure 17.4 shows typical response
curves of yield against NPK applied. These data are from experiments
on magnesian limestone soils at N index 0, P index 3/4 and K index 2/3.

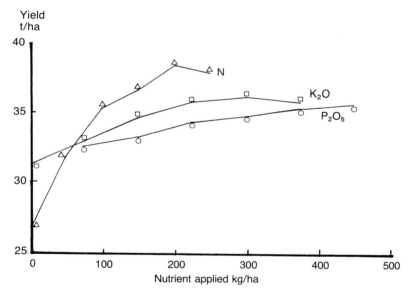

Figure 17.4. NPK response of maincrop potatoes
Source: ADAS

The very shallow phosphate response curve is typical. The economic optimum is very difficult to define between 200 and 400 kg/ha P_2O_5. Hence the reason for differences of opinion on how much phosphate should be recommended. The main conclusion is that at index 3 or above at which most potatoes are grown, yield differences are very small for a wide range of application rates. If a generous application is made, it is not necessary to apply phosphate for other crops in the rotation.

Potatoes are very responsive to fresh phosphate fertiliser and a minimum of 100 kg/ha P_2O_5 is recommended however high the soil P index. Experiments have shown that autumn application is not satisfactory. All the phosphate should be applied at or just before planting. Ideally broadcast applications should be worked in during seedbed cultivations. Where stone separation is carried out the fertiliser should be broadcast first. Only materials with at least 60 per cent water-soluble phosphate should be used for potatoes. At low soil indices, a higher water-soluble content is preferable.

Phosphate response is usually associated with an increase in number of ware sized tubers rather than just increasing tuber size. However, any difference is small compared to seed spacing effects on tuber number.

Potassium

Potatoes respond to potassium particularly at low soil indices. At index 1 very large yield increases are found. At index 4 or above the response is small. Figure 17.4 shows the typical response at soil levels of 200–300 mg/l K (upper index 2/low index 3). It is common for yield to be slightly reduced when more than 300 kg/ha K_2O is broadcast before planting, in addition to the nitrogen. This is almost certainly due to excessive soluble fertiliser causing damage to young roots and shoots.

Work at Norfolk Agricultural Station has shown a better response if part or all of the potash is applied in autumn or at least six weeks before planting, allowing time for the chloride to leach. This work was carried out on a K index 1 soil. As shown in Table 17.7, yield continued to increase and a greater yield response was shown at lower application rates. Where high rates of potash are used, part should be applied in autumn or winter, but 100 kg/ha K_2O is probably best applied near planting. Potash yield benefits are mainly due to larger tubers.

Method of Application

Placement of part or all of the fertiliser application in two bands just below and to the side of the seed at planting is common commercial practice. The Rothamsted experimental data in Figure 17.5 show that

**Table 17.7. Effect of rate and timing of potash on yield of maincrop potatoes
(1976, 78 and 79, K index 1)**

Timing	kg/ha K$_2$O			
	0	200	400	600
		Yield (t/ha)		
Autumn	25.0	31.0	33.2	34.8
6 weeks before planting	24.5	31.8	33.5	34.5
Seedbed	23.4	29.7	30.4	31.5
100 kg/ha in seedbed plus rest 6 weeks before planting	23.1	31.2	34.3	35.6

Source: Norfolk Agricultural Station.

the main yield benefit from placement occurs at low application rates:
it is a disadvantage to place too much fertiliser. Recent work by Norsk
Hydro has shown that if only part of the fertiliser is placed a small yield
benefit above that given by all broadcast can be achieved. At the rates
of fertiliser now recommended it is no longer appropriate to think in

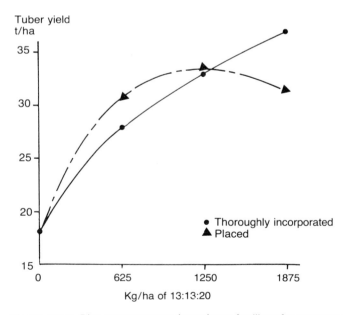

Figure 17.5. Placement versus broadcast fertiliser for potatoes
Source: Rothamsted Experimental Station

terms of saving fertiliser by using placement. No more than 250 kg/ha $N + K_2O$ should be placed to avoid adverse effects on the crop. If part of the application is placed, the same total nitrogen and potash rates should be applied as if all broadcast. A small reduction in phosphate use may be made if most of the phosphate is placed. A more practical option is usually to apply a generous rate of phosphate for the potatoes and omit phosphate elsewhere in the rotation. Reductions in yield from too much fertiliser near the seed whether placed or broadcast are greater with chitted seed and when planted into dry soil.

Tuber Quality
Comment has already been made regarding NPK effects on tuber size. Increasing rates of both nitrogen and potash tend to reduce tuber dry matter content or specific gravity. Phosphate has no effect. An increase in nitrogen rate of 100 kg/ha N will commonly decrease the crop percentage DM by about 0.5–0.7 per cent. A similar increase in use of potash will produce a rather smaller decrease in percentage DM. Potassium sulphate (sulphate of potash) produces tubers of higher dry matter content than potassium chloride (muriate of potash) for comparable potash application rates. The difference is likely to be up to 1 per cent DM at normal application rates.

High nitrogen delays maturity. This is mainly of importance in earlies and sometimes second earlies. The delay is due mainly to increased green leaf duration. It can be a problem in late maturing varieties. High nitrogen can also increase the incidence of after-cook blackening in susceptible varieties. It is more common in crops grown on peat soils.

Internal Bruising (Blackspot)
Susceptibility to internal bruising in a particular variety depends on several factors including both dry matter content and tuber potassium level. High dry matter varieties such as Record are very susceptible. In recent experiments Pentland Crown and Desirée have both shown reduced internal bruising susceptibility related to high potash fertiliser application. Work in England and Holland has shown that internal bruising levels are low in tubers with 2.5 per cent K or more (DM basis). Table 17.8 shows data collected from a survey of field crops in the East Midlands in 1973. Susceptibility was assessed using a laboratory falling bolt technique. Large applications of potassium chloride will both decrease dry matter and increase tuber K content. Each contributes to a lower internal bruising susceptibility. Tubers grown at low soil K index have a lower K content than those grown at index 2–3.

Table 17.8. Effect of tuber potassium content on susceptibility to internal bruising

| | Per cent K (DM) | | | |
| | <1.9 | 1.9–2.2 | 2.2–2.5 | >2.5 |
Variety	Per cent tubers affected			
Record	27.2 (12)	19.6 (17)	15.5 (12)	4.2 (4)
Pentland Crown	6.1 (6)	3.3 (14)	1.0 (19)	0.0 (7)
Pentland Ivory	6.0 (5)	3.3 (9)	1.4 (12)	0.5 (6)

Note: Number of samples in parentheses.
Source: ADAS.

Potatoes for Processing
The particular quality requirements for the various methods of potato processing may influence fertiliser use. For crisps, the dominant variety Record is chosen for its high dry matter content. A small increase in dry matter content can be achieved by using sulphate rather than chloride potash fertiliser but internal bruising may increase. Choice will depend on whether high dry matter or reduced bruising offers the greater financial attraction. Excess nitrogen should not be used. Work at Gleadthorpe EHF has shown adverse effects of organic manures on crisp quality. This is probably due mainly to a reduction in tuber percentage dry matter caused by the higher nutrient levels. Fertiliser use is unlikely to be varied for crops grown for french fries or for dehydration. Recent work has shown that nitrogen rate has no effect on crop use for french fry production. If crops are grown specifically for canning, the fertiliser rates recommended for earlies are appropriate.

Early Potatoes
First early varieties grown for lifting in May or June require less nitrogen and potash than maincrops. This is mainly due to lower yield and therefore nutrient uptake. At N index 0, 180 kg/ha N is adequate. For very early lifts, 150 kg/ha N is appropriate for yields up to 15 t/ha. Too much nitrogen will delay tuber initiation and therefore lifting date. Phosphate response is at least as great as for main crops and higher rates are generally recommended, particularly at index 3–4. At index 2 for potash, 120 kg/ha K_2O is sufficient.

Care should be taken with crops grown under polythene. The natural drying out of the soil that occurs before the sheet is removed makes the crop susceptible to excess soluble fertiliser in the soil. This can stunt growth. The early lift nitrogen rate and the early potash application rate should not be exceeded. Overlaps during fertiliser application must also be avoided.

Seed Potatoes

Where crops are grown specifically for seed and burnt off early 180 kg/ha N is adequate. Phosphate and potash is required as for ware crops.

Magnesium

Magnesium deficiency symptoms are common in the older leaves of senescing potato plants (see Plate C6). Application of magnesium fertiliser is rarely of benefit in these situations. On soils at Mg index 0 or 1 magnesium should be applied at 100 and 50 kg/ha Mg respectively. A yield response is only likely at index 0, but an insurance application is appropriate at index 1. Sometimes on low magnesium soils excessive potash use may induce magnesium deficiency symptoms.

Manganese

Potatoes sometimes suffer from manganese deficiency when grown on organic soils with a pH above 6.0. Deficiency is not common but should be treated by foliar spraying if symptoms are seen. Potatoes are one of the few crops which exhibit symptoms on the younger leaves.

Chapter 18

OILSEED RAPE AND OTHER ARABLE CROPS

THIS CHAPTER covers winter and spring oilseed rape, field beans, dried peas and herbage seed. Brief mention is also made of a number of other brassicas and minor arable crops. Most of the latter have had very little experimental work undertaken on their fertiliser requirements in the UK.

OILSEED RAPE

The expansion in oilseed rape growing in the UK in the last ten years has encouraged considerable experimental work by ADAS and many others on various aspects of production. A considerable number of experiments have looked at the fertiliser requirements of the winter crop. Current experimental work is looking at a restricted range of aspects in more detail; the broad picture of NPK requirement under UK conditions has been established.

Acidity
Oilseed rape is moderately susceptible to acidity. A normal arable pH of 6.5 on mineral soils and 5.8 on organic soils is recommended. Acidity symptoms will show on mineral soils below about 5.8. If a pH below this is found after establishment, topdressing with ground chalk or limestone is recommended.

Nitrogen for Winter Oilseed Rape
The uptake of nitrogen during the autumn varies considerably depending mainly on the date of emergence. Crops emerging in August may contain 80 kg/ha N by the end of November; crops emerging a month later will often contain less than half this quantity. Most crops show a loss of older leaves during late autumn and winter, with some nitrogen returned to the soil. The main nitrogen uptake

period is from mid April to mid June. Most crops will reach a maximum uptake in June of 200–250 kg/ha N. This total will fall as the crop ripens and loses leaves during July. The amount of nitrogen removed in the seed is about 33 kg/t at 92 per cent DM.

Response to Spring Nitrogen
Most crops show a response to spring topdressing in the range 4–8 kg seed/kg N applied. The economic breakeven rate is only just over 1 kg seed/kg N. Thus an extra 50 kg/ha N only needs to produce an extra 60 kg/ha of seed to break even. In most experiments the economic optimum nitrogen rate is very close to the rate giving maximum yield.

Table 18.1. Effect of winter oilseed rape yield on response to nitrogen topdressing

Yield level at optimum N (92 per cent DM) t/ha	Number of sites	Optimum nitrogen kg/ha
Less than 2.5	6	140
2.5–3.5	18	196
Greater than 3.5	12	236

Source: ADAS.

Experiments on spring nitrogen topdressing have been carried out throughout the main rape-growing area of England, from Yorkshire to Wiltshire. Economic optimum N rates for individual sites varied from 75 to 325 kg/ha N. ADAS results have been examined to find the major factors affecting nitrogen requirement. Previous cropping was always a cereal and there were no consistent effects of soil type, soil organic matter content or sowing date. The main factor responsible for the variation in optima was the crop yield. Table 18.1 gives the median optimum nitrogen rate at different yield levels.

Analysis of these experiments indicates that 200 kg/ha N is likely to be adequate on sandy soils, because yield levels are unlikely to exceed 3.5 t/ha. On potentially high yielding sites where experience suggests that yields greater than 3.5 t/ha are probable, 240 kg/ha N should be applied. This yield level is most likely on uniform early established crops on medium or heavy soils, with no serious pest damage or weed competition.

Timing of Spring Nitrogen
Work at Sutton Bonington and elsewhere has shown that yield potential depends on pre-flowering growth and dry matter production.

The greater the crop growth during March and April, the higher the yield potential. For this reason it is important that the crop has adequate nitrogen during this period, particularly at the start of spring growth.

ADAS carried out twenty-four experiments during the 1970s in which different times of application were tested at two different rates of application, 150 and 225 kg/ha N. The early dressing was applied when spring growth started in late February to mid March. The late

Table 18.2. Effect of timing of nitrogen topdressing on yield of winter oilseed rape

Timing	Yield t/ha
All early	2.86
50 per cent early 50 per cent late	2.85
25 per cent early 75 per cent late	2.76

Source: ADAS.

application went on about four weeks later in late March to mid April. The results are summarised in Table 18.2. In each year all early and 50:50 split gave very similar yields, while the 25:75 split gave the lowest average yield in each year. There was no effect of soil type on the best timing.

These results confirm current advice that a single application at the start of spring growth in the last week of February or early March is the most suitable timing on medium and heavy soils. If farm organisation means a 50:50 split is preferable, no loss of yield is likely. All the nitrogen should be applied by the end of the first week in April. If soil wetness prevents early ground application, 50 kg/ha N should be applied by air followed up by the remainder as soon as the soil is dry enough. As long as 25 per cent is applied early, the yield penalty is likely to be small.

On sandy soils and shallow soils over chalk or limestone where the leaching risk is high in early spring, a 50:50 split is preferable. This is particularly important if spring growth on these soils starts in mid February, as may occur in southern England. The remainder of the nitrogen should be applied by the end of the first week in April.

Experiments carried out recently on heavy land in East Anglia showed no advantage in yield from early February application. The treatments tested were 0, 50 or 100 kg/ha N in early February and the remainder in early March.

Autumn Nitrogen

The yield response to seedbed nitrogen is much less than to nitrogen applied in the spring, Research work has shown that it is important to achieve an adequate plant size in the autumn before inflorescence initiation, which occurs from December onwards depending on conditions. Plant size will depend mainly on date of emergence. Autumn nitrogen will generally produce an increase in the size of plants, by

Table 18.3. Effect of winter oilseed rape yield on response to seedbed nitrogen

Yield level (92 per cent DM) t/ha	Number of sites	Number responding to autumn N
Less than 2.5	8	0
2.5–3.5	12	2
Greater than 3.5	8	6

Source: ADAS.

increasing leaf area during early growth. The difference is most noticeable in early-emerging crops. Examination of the results from twenty-eight ADAS experiments (Table 18.3), in which autumn nitrogen rates were included, shows no overall yield benefit. On average over these sites the yield benefits from autumn nitrogen did not cover the cost of the fertiliser. There was no indication that autumn nitrogen affected response to spring topdressing. However eight individual sites out of the twenty-eight did give yield responses large enough to cover the fertiliser cost. A more detailed breakdown of the sites on the basis of yield level showed this to be related to response. Soil type was not an important factor.

The higher yielding sites were the ones most likely to respond to autumn nitrogen, sufficient to cover the cost of the fertiliser. No responses above 50 kg/ha N were found. In practice autumn nitrogen at 50 kg/ha N is recommended, but yield benefit is small and largely unpredictable at the time of application. Many farmers prefer as large a plant as possible during the winter to minimise the risk of pigeon damage. Rates above 50 kg/ha N are unlikely to give any further increase in plant size.

Oil Content

Autumn-applied nitrogen has no appreciable effect on oil content. Spring nitrogen reduces oil content with increasing rate of application. This is balanced by an increase in protein content. Experiments show a linear reduction in oil content over the range 75–325 kg/ha N topdressing. The decrease was 1.25 per cent oil (DM basis) for every 100 kg/ha N applied. This decrease is not large enough to have a significant effect on the economic optimum rate of nitrogen topdressing.

Nitrogen for Spring Oilseed Rape

Much less work has been done on spring than on winter oilseed rape. Results of work carried out in the mid 1970s suggest that the crop requires slightly less nitrogen. In the normal rotational position following a cereal, 150 kg/ha N is recommended. On sandy or dry seedbeds only 50 kg/ha N should be applied before drilling with the remainder topdressed after emergence. This avoids the leaching risk and possible damage to young seedlings.

Phosphate

Seedbed phosphate rates were tested in twenty-two ADAS sites. Overall the yield response to phosphate was just adequate to cover the cost of the fertiliser. Table 18.4 shows the breakdown for different soil

Table 18.4. Winter oilseed rape response to phosphate

Soil index	Number of sites	Yield response (t/ha) 25 kg/ha P_2O_5	50 kg/ha P_2O_5
0–1	8	+0.13	+0.19
2	7	+0.12	+0.10
3+	7	−0.12	−0.07

Source: ADAS.

analysis indices. The worthwhile yield responses were achieved at index 0–1. The negative effects at index 3+ were small and variable. No sites showed a worthwhile benefit to rates above 50 kg/ha P_2O_5.

Current recommendations are 50 kg/ha P_2O_5 at indices 1–3, to ensure optimum yield response and maintain soil reserves. Offtake of phosphorus as P_2O_5 in the seed is about 15 kg/t. Thus 50 kg/ha P_2O_5 is adequate for 2.5 t/ha but should be raised to 60 kg/ha P_2O_5 where crop yields average 4 t/ha. At least 75 kg/ha P_2O_5 should be applied on index 0 fields.

Potash

The twenty-two phosphate sites also tested potash response (Table 18.5). Overall there was no yield response to potash and no individual site responses that could be predicted from soil analysis or soil type. This was despite the fact that nine of the twenty-two sites were in soil index 1 for potassium. Potash should be applied to maintain soil levels on all soils at index 1 and on sandy soils at index 1 or 2. Offtake in the seed as K_2O is about 11 kg/t. The current recommendation of 40 kg/ha K_2O is adequate for maintenance situations, although at least 75 kg/ha K_2O should be applied on index 0 fields.

Table 18.5. Winter oilseed rape response to potash

Soil index	Number of sites	Yield response (t/ha)	
		25 kg/ha K_2O	50 kg/ha K_2O
1	9	−0.06	+0.02
2+	13	−0.05	−0.01

Source: ADAS.

Magnesium

Magnesium deficiency symptoms are common on the older leaves of maturing plants as seen in many other crops. In the absence of experimental data on yield response, magnesium is recommended at soil index 0 only. It is very unlikely that any yield benefit will be obtained at higher soil levels.

Sulphur

Recent experiments in the low atmospheric sulphur deposition areas of south-west England have shown very small but statistically significant and economically worthwhile responses to sulphur in winter oilseed rape. This result was obtained on a few sites on thin chalk soils. The need for sulphur fertiliser use on the crop is still very localised in England and Wales. It remains to be seen whether soil analysis can be used to predict responsive sites.

Boron

Work in the last few years has shown yield responses to boron on sandy soils, where soil boron levels are low. Yield benefits have been obtained without any obvious symptoms of boron deficiency in the crop. At the present state of knowledge, boron should only be applied on soils with a soil boron level of 0.5 mg/l available boron or below. This may be applied as two foliar sprays; one should be applied in the autumn and one in the spring.

Manganese
Manganese deficiency is rarely seen in oilseed rape on mineral soils. Alkaline sands are the soils most likely to show the problem. Deficiency is seen from time to time on alkaline organic soils (see Plate C14). Treatment is a foliar spray of manganese sulphate at 9 kg/ha plus wetter in at least 250 t/ha of water.

HERBAGE SEEDS

Although a large number of herbage species are grown for seed production in the UK the majority of the area is in ryegrasses. The only recent fertiliser work has been on the newer perennial ryegrasses. The fertiliser requirements for grass establishment are the same as given in Chapter 15.

Nitrogen
The recent work has shown an economic benefit from 150 kg/ha N in the production year on sandy and shallow soils over chalk. On other mineral soils, 120 kg/ha N is appropriate. The recommendations for other grasses are 170 and 120 kg/ha N respectively for cocksfoot and 100 and 80 kg/ha N respectively for timothy, tall and meadow fescue. These recommendations are for spring topdressings. If too much nitrogen is given, secondary tillering and early lodging can cause harvesting difficulties. The recommendations should not be exceeded. They assume a previous cereal rotation. Where a single application is used, it should be applied in March. Recent work by National Institute of Agricultural Botany (NIAB) and ADAS in Hampshire has shown some advantage from a split application on early perennial ryegrass. Half was applied in mid March and the remainder in mid April. Application of 50 kg/ha N immediately after seed harvest is recommended for crops to be grazed or cut for conservation in the autumn.

Phosphate and Potash
The same recommendations as for hay in Chapter 15 are appropriate.

DRIED PEAS

Little experimental work on dried peas has been carried out in the UK since the early 1950s. There is little reason to doubt the basic findings of those studies in relation to current circumstances.

Nitrogen

There has been considerable debate over the years as to the benefits of small seedbed applications of nitrogen for early drilled pea crops, particularly in wet springs when nodulation may be poor. While yield increases were shown in some early work, yield depressions were also shown. On balance the recommendation is not to apply seedbed nitrogen. Soil nitrogen supply will be adequate to provide any early requirement.

Phosphate and Potash

Early experiments emphasised the need for potash and benefits from placement of nutrients below the seed. It also showed that combine drilled fertiliser could easily cause damage and reduce plant population. These experiments were often on soils with much lower soil P and K indices than are common today. At index 2 or above small maintenance applications of P and K are required. As the crop requires no nitrogen it is common practice on many farms to omit all fertiliser for this crop and make up the PK balance elsewhere in the rotation. Yield response is only likely at index 0 and perhaps index 1 for potash. At index 0, 150 kg/ha of each of P_2O_5 or K_2O is needed. Care should be taken to avoid too much potash fertiliser near the seed as the crop is sensitive to excess soluble nutrient. High rates should be well cultivated in or applied earlier so the chloride can leach below sowing depth. Placement of fertiliser for peas is unlikely to be used in practice as suitable machinery is not available and not justified on most soils.

Magnesium

Visual symptoms of magnesium deficiency in the crop are commonly due to poor root activity, often induced by waterlogging and associated with poor nitrogen fixation and uptake. Yield response to magnesium fertiliser is only likely on low index 0 fields.

Manganese

Marsh spot of peas was investigated in the 1950s, particularly in Holland. The symptom of internal breakdown of the seed is only visible at harvest. It may or may not be associated with leaf symptoms earlier in the life of the crop. A low level of incidence will cause crop rejection for some markets, so insurance treatment with manganese is recommended on soils susceptible to deficiency. These are organic soils above pH 6.0 and sandy soils above 6.5.

The rate of application is usually best limited to 5 kg/ha of manganese sulphate in at least 250 l/ha of water plus wetter. Crop scorch may occur if a higher concentration is applied. As manganese movement in the plant is poor, timing of application is important. The best timing is when the flowers are fully open. A two-spray programme is recommended at early flowering and ten days later on sites with a history of manganese deficiency. If a spray is applied before flowering because leaf symptoms are seen in the crop, the marsh spot control programme should still be applied.

FIELD BEANS

Less work has been carried out on field beans than on dried peas. Occasional experiments have confirmed that fertiliser nitrogen to supplement or replace that fixed by the crop gives no yield advantage.

Phosphate and Potash

Like peas, field beans are unresponsive to phosphate and potash except at very low soil levels. At soil index 0 for P or K, 150 kg/ha P_2O_5 or K_2O should be applied. Otherwise maintenance applications are adequate. A 4 t/ha crop will remove about 45 kg of P_2O_5 and 50 kg of K_2O. The crop is sensitive to excess soluble fertiliser in the seedbed. The large potash requirement at index 0 should be thoroughly cultivated in before drilling. Combine drilling is best avoided.

MUSTARD FOR SEED

Work by Colmans has shown that the crop responds to similar rates of nitrogen as oilseed rape. The following rates are recommended—200 kg/ha N on light soils, 160 kg/ha N on medium and heavy soils and 100 kg/ha N on organic soils. Only part of the recommendation should be applied in the seedbed on sandy soils. Phosphate and potash should be applied as for oilseed rape.

OTHER BRASSICA SEED CROPS

Summer-sown crops of kale, swedes or turnips for seed production have attracted no experimental work on fertiliser need. In general, autumn nitrogen is not applied, particularly if a variety is known to be susceptible to winter kill. Early spring topdressing is usual at 120–150 kg/ha N. Application is best done by air or high clearance machinery. Where the crop is susceptible to lodging, this will be aggravated by high rates of nitrogen topdressing. Even application is essential to ensure a uniform crop with a high germination percentage at harvest. Apply phosphate and potash as for oilseed rape.

LINSEED

A few experiments in the 1970s showed that following a cereal crop an appropriate nitrogen rate is 120 kg/ha N on light soils and 80 kg/ha N on medium and heavy soils. Phosphate and potash should be used as for oilseed rape.

Chapter 19

VEGETABLES AND BULBS

OVER THIRTY different vegetable and bulb crops are grown on a significant scale in the UK; however, only the more important ones are covered in this chapter. Many of these have little experimental work on which to base their fertiliser requirements. For this reason, work at National Vegetable Research Station (NVRS), Wellesbourne, has been aimed at modelling the NPK response of vegetable crops. Each crop is characterised by limited experimentation and its fertiliser requirement is then derived from comparison with crops such as sugar beet and potatoes on which many experiments have been carried out on a wide range of soil types. These results are then able to be tested and modified as appropriate using the response experimental data available on some of the more important vegetable crops. The outcome has been useful for phosphate and potash recommendations, but nitrogen response is still largely derived from empirical experimentation.

VEGETABLES

Vegetables are grown in two different farming systems in the UK. A wide range of vegetable crops is grown on intensive holdings in the Thames Valley, the Vale of Evesham and elsewhere, generally on light loamy soils. These holdings have predominantly vegetable rotations and commonly have soil P indices of 5 or above and soil K indices of 3 or 4. Crop response at these levels is unlikely for any vegetables. Application of phosphate and potash is only needed on a maintenance basis. Often phosphate is not justified at all.

Vegetables which are grown on a larger scale and can be fully mechanised are found mainly in arable rotations, often being the only vegetable crop on the farm. Production is concentrated on suitable soil types or in suitable climatic areas. Examples are carrots on sandy soils in Norfolk and beetroot on silts and organic sands in Humberside. The less hardy types of winter cauliflower are limited to coastal areas of south west England and Wales. Winter celery is grown on peaty soils in the Fens and Lancashire. Swedes are important in south Devon. Vegetables grown for freezing, particularly broad beans, french beans, vining peas and sprouts, are limited to areas near to the freezing-factories. These are almost all in eastern England.

While the soil PK index in these situations is lower and perhaps more variable than those on intensive vegetable holdings, P index will rarely be below 3 while a K index of 2 will be normal, except for sandy soils at index 1.

Liming

Acidity is a common problem of many vegetable growing areas. Lettuce and celery are the most sensitive to low pH. Most vegetables require a pH of at least 6.0 to grow satisfactorily. Detailed checking with pH indicator for fields intended for a sensitive vegetable crop is essential on potentially acid soils. For all vegetables, the pH of mineral soils should be 6.5 or above (5.8 on peats). If lime is needed, the rate must be correct for the ploughing depth which will generally be greater than the standard 15 or 20 cm on which lime recommendations are calculated. Where brassicas are grown frequently and club root is a serious risk, the pH should be kept at 7.0 on light and heavy loams. If this is attempted on sands, the risk of both manganese and boron deficiencies in susceptible crops will need to be taken into account. Sensitive crops should not be grown immediately after liming very acid soils, as the pH may not be raised quickly enough.

Crop Establishment

Many vegetable crops are direct sown in the field. For most a uniform and precise plant population is one of the most important aspects of producing the required quantity and uniformity of crop. This is also one of the most difficult to achieve. If the water supply to the very young seedling is marginal, the higher the concentration of soluble fertiliser in the seedbed, the greater will be the adverse effect on the plant population. Any increase in the soil water osmotic concentration will be a disadvantage.

Work at NVRS, Wellesbourne, has shown that nitrogen fertiliser should only be used in small quantities so that the seed bed nitrate content is low until crop establishment is complete. Some is necessary or the initial growth rate will be too slow. Total crop requirement at this stage of crop growth is low. High nitrate levels in the soil before the crop root system has developed are at risk from being leached below root depth by rainfall or irrigation. For many crops the majority of the nitrogen requirement should be applied after full establishment, with only 50 kg/ha N applied in the seedbed. Bare rooted transplants such as leeks or brassicas can also be damaged by too much nitrate in a dry seedbed and growth can be delayed, although loss of plants is unlikely. The maximum application for these crops is best restricted to 100 kg/ha N before planting. The remainder can be topdressed once the plants are growing away. Table 19.1 shows typical effects of

Table 19.1. Effect of seedbed ammonium nitrate on establishment of vegetable crops (expressed as percentage of nil nitrogen treatment)

	N applied (kg/ha)				
	0	*60*	*120*	*180*	*240*
Lettuce	100	89	76	54	51
Carrots	100	97	87	75	67
Redbeet	100	95	86	75	79
Onions	100	93	76	64	65

Source: National Vegetable Research Station.

seedbed ammonium nitrate application rates on plant population. Lettuce and onion were more sensitive than carrots or redbeet, but the establishment of all was reduced.

Other fertilisers that will add significantly to the total osmotic concentration in the seedbed are potassium chloride and salt. The main cause of trouble is the chloride in these fertilisers. Salt should always be applied several weeks before drilling so that the chloride is leached down the profile by winter rainfall. Modest potash rates can be tolerated in the seedbed but are better applied the previous autumn at K index 2 or above. The main difficulty is dealing with index 1 sandy soils, for here potash yield response is greatest but osmotic effects are most damaging. In these circumstances application after ploughing but a few weeks before drilling is the ideal. Alternatively where a high rate of potash is needed, some can be autumn applied with the remainder in the seedbed.

Nitrogen

The major vegetable crops have an experimental basis on which to make fertiliser recommendations. However it is usually restricted to the main soils and principal areas of the country in which the crop is grown. Rarely are data available on different rotational positions and their variation in soil nitrogen supply. Most of the experimental work has been on crops following cereals rather than in more intensive rotations. Another problem is the high experimental error common in most vegetable experiments. Some is due to variation in plant population as commonly seen in root crops. Brassicas often show marked plant variability even with F_1 hybrids. It is often difficult to show a well-defined response curve and economic optimum nitrogen rate. The less important crops rely on a mixture of a few experiments, field observation and experience.

The nitrogen recommendations in Table 19.2 are derived from ADAS publications and are suitable for light loamy soils. Unless stated, 20 per cent higher rates will generally be appropriate on sandy and shallow soils over rock and 20 per cent less on deep silty soils. Specific autumn and spring topdressing recommendations for over-wintered crops are given. Where crops are commonly grown on organic soils special recommendations are made. For many spring and summer sown crops, only part of the nitrogen should be put in the seedbed to avoid adverse effects on plant establishment. The remainder should be topdressed after the crop is well established.

Some vegetable crops can be damaged by topdressed nitrogen whether as prills or liquid. Damage is worse if small prills or dust sticks to the leaf or liquid is applied with too many fine droplets. The risk of damage can be minimised by a good product and the right application technique. The main problem with prills is lodging in leaf axils and damaging tissue which may subsequently be invaded by disease. For some crops such as sweetcorn severe damage may be caused if growth is too far advanced before the topdressing is applied. Other crops such as lettuce are easily damaged by liquids. Early spring topdressing during frosty weather gives a greater risk of leaf scorch. Topdressing just before or during rain is one way of avoiding scorch from liquids or solids if a crop is particularly at risk.

Phosphate and Potash

Work at NVRS has shown that most vegetables give only a very small yield response to phosphate at soil index 3 or above and potash index 2 or above. The exceptions are lettuce which responds at higher soil P levels and spinach which responds at higher soil levels of both P and K. More generous applications of phosphate than determined by offtake are needed for these crops at P index 3.

Most vegetable crops require either a maintenance application of phosphate or none at all if the soil index is very high. Indices 2 and 3 should be maintained, but application at index 4 is difficult to justify. At index 5, phosphate is not needed.

Table 19.2. Nitrogen recommendations for vegetable crops

	ADAS nitrogen index		
	0	*1*	*2*
Crop	*kg/ha N*		
Asparagus – before planting	150	75	50
– 2nd year	100	100	100
– 3rd year onwards	125	125	125
Beans – broad	60	25	0
– french	150	100	75
– runner	150	100	75
Beetroot	250	200	150
Brussels sprouts	300	250	200
Cabbage–summer/autumn	300	250	200
– winter/savoy	300	250	200
post-Christmas cutting	150	125	100
topdressing	75	75	75
– Dutch white for storage	250	200	150
– spring	75	50	25
topdressing – greens	100–200	100–200	100–200
– hearted	400	400	400
– early frame raised	250	200	125
topdressing	60–120	60–120	60–120
Calabrese	250	200	160
Carrots –	60	25	0
fen peats	0	0	0
Cauliflower – summer and autumn	250	200	125
– winter Roscoff types	75	40	0
topdressing	60–125	60–125	60–125
– winter hardy	75	40	0
topdressing	125–200	125–200	125–200
Celery – winter – peats	0	0	0
– peaty loams	50	50	50
– moss peats	100	100	100
– self-blanching	75	75	75
topdressing	75–150	75–150	75–150
Courgette	100	75	50
topdressing	75	75	75
Leeks – mineral soils	150	100	60
topdressing	100	100	100
– fen peats	60	30	0
Lettuce – summer	125	100	75
– sandy soils	200	150	100
– over wintered/topdressing	75–150	75–150	75–150
Marrow –	100	75	50
topdressing	75	75	75
Mint – preplanting	125	100	50
– spring and after each cut	75	75	75

Table 19.2. (continued)

| Crop | ADAS nitrogen index | | |
| | 0 | 1 | 2 |
	kg/ha N		
Onion – salad – summer/autumn	125	75	50
– salad – winter	25	0	0
topdressing	50–125	50–125	50–125
– bulb spring sown	90	60	30
– fen peats	30	0	0
– bulb autumn sown	50	25	0
topdressing	100	100	100
– fen peats	0	0	0
topdressing	50	50	50
Parsley – seedbed	75	50	25
– after each cut	50	50	50
Parsnip – mineral soils	100	75	0
– fen peats	60	40	0
Peas – vining/green	0	0	0
Radish – 1st crop	60	25	0
– succeeding crops	40	0	0
Rhubarb – establishment	175	125	75
– field crop annually	250	250	250
Spinach – summer	150	100	50
topdressing	200	200	200
– over-wintered	0	0	0
topdressing	125	125	125
Sweetcorn	100	75	50
Swede	100	50	0
Tomato – outdoor	100	50	50
Turnip – maincrop	100	50	0
– early bunching	150	100	50

Source: ADAS.

Crop removal of potash is high for many vegetable crops and this should be replaced at indices 1 and 2. At index 3 half maintenance is appropriate for high offtake crops. No potash is needed at index 4 or above.

Generous applications of both phosphate and potash are recommended before planting asparagus or rhubarb. The figures in Table 19.3 are calculated to balance the phosphate and potash removed in average crops of the more important vegetables. They provide a basis for ensuring that soil levels of P and K are maintained.

Table 19.3. Maintenance fertiliser levels of phosphate and potash for the main vegetable crops

Crop	P_2O_5 kg/ha	K_2O
Beans – broad	10	20
– french	10	20
Beetroot	40	160
Brussels sprouts – stems ploughed in	35	80
– stems removed	60	160
Cabbage	30	110
Carrot	30	110
Cauliflower	40	150
Lettuce	20	100
Onion – bulb	20	60
Parsnip	60	150
Peas	10	20
Swede	30	100
Turnip	30	100

Source: ADAS.

Magnesium

The experimental evidence of vegetable crop response to magnesium is sparse. Many show magnesium deficiency symptoms which may be due to a low soil magnesium level, particularly on sandy soils. At index 0, apply 100 kg/ha Mg every third year in intensive vegetable rotations or for each vegetable crop in arable rotations. At index 1 apply 50 kg/ha Mg in the same way.

Sodium

There is some uncertainty over which vegetable crops will benefit from sodium in addition to adequate potassium. Fen peats and silts usually contain adequate sodium, so vegetable crops on these soils do not justify salt application. The biggest yield responses are shown on sandy soils. Carrots justify 150 kg/ha Na (400 kg/ha salt) when grown on these soils. It must be applied several weeks before drilling to avoid adverse effects on crop establishment. Other vegetable crops that justify sodium application when grown on sandy soils are beetroot and celery.

Trace Elements

While a number of trace element deficiencies have been diagnosed in vegetable crops in the UK, few justify routine treatment on an insurance basis. Nevertheless it is important to ensure that treatment is applied where the whole crop risks being unmarketable if only slightly affected by a trace element deficiency.

Boron
Deficiency occurs on sandy soils, usually above pH 7.0. Soil analysis is a useful guide to the risk of deficiency. As boron is easily leached, each susceptible crop must be treated in low boron fields. Most sandy fields are below 1 mg/l soil-available boron and the following crops grown in this situation should receive boron to ensure that crop quality is not spoilt by deficiency symptoms: carrots, swedes, beetroot, celery, cauliflower.

Manganese
Manganese deficiency is less damaging than boron, but more common in occurrence. Most crops can show symptoms when grown on susceptible soils and should be treated as soon as symptoms appear. Some sandy and organic fields at high pH will need treatment every year.

Molybdenum
Cauliflower and lettuce are the only crops likely to show field symptoms of deficiency. Liming to pH 6.5 usually avoids the problem. Symptoms are more common during plant raising of these crops in peat composts, and routine treatment with molybdenum during this stage is important.

BULBS

Outdoor bulbs are grown predominantly in eastern England for dry bulb production. Narcissi and tulips are dominant, but a few irises and gladioli are also grown. Narcissi are also important in Cornwall and the Isles of Scilly where they are grown primarily for early flower production. Anemones are also important in these areas.

Bulbs are different from most other crops in their nutrient response. As the majority of the crop yield is due to an increase in size of each of the bulbs planted, the nutrient status of the bulb stock has a greater influence on fertiliser response than is shown by the very small effect of seed nutrient content of most other crop plants. The effect of nutrient supply is more easily assessed by its effect on maintaining a satisfactory bulb nutrient content than on yield response in an annual experiment. Bulb nutrient content is especially important in bulbs grown for forced flower production. To show differences in flower production and quality in the field, it is usually necessary to run down the bulb nutrient content for three or four production cycles by not applying nitrogen or other nutrients on soils at index 0 for the particular nutrient.

Liming
Narcissi will produce satisfactorily at pH 5.5 but tulips and other bulbs generally require a pH of 6.5.

Nitrogen
Some experimental work has been carried out on narcissi and tulips in the UK. For the tulip crop planted in October, recent ADAS experiments have shown no effect of autumn nitrogen on bulb nitrogen content or yield. Nitrogen topdressed in February at the one-leaf emerged stage increased yield and bulb nitrogen content as shown in Table 19.4, but decreased bulb dry matter content. The result is a

Table 19.4. Effect of nitrogen topdressing on tulips

	N applied (kg/ha)				
	0	37.5	75	112.5	150
Bulb yield (> 10 cm)	4.68	4.65	4.88	4.87	5.07
Bulb % DM	42.90	42.10	41.30	41.00	40.60
Bulb % N	1.08	1.34	1.44	1.53	1.70

Source: ADAS.

recommendation of 100 kg/ha N applied in February on N index 0 sites. This will maintain bulb nitrogen content at around 1.5 per cent N yet produce a good quality bulb with a satisfactory dry matter content. On silty soils or at N index 1, 50 kg/ha N should not be exceeded.

The normal two year narcissus crop is planted in August and produces a greater yield increase than tulips over the weight of bulbs planted. Although nitrogen is sometimes recommended at planting, it is doubtful if the crop benefits. A spring application of 50 kg/ha N in each year on sandy soils is probably better. On silty soils very little benefit from applied nitrogen is likely.

Little experimental evidence exists for the other crops. An application of 75 kg/ha N at planting has been shown adequate for anemones. Iris should receive a similar rate at planting, while gladioli are more responsive and can benefit from nitrogen topdressing.

Other Nutrients

Unless grown at a very low soil index, only modest applications are needed to balance offtake and maintain soil levels. For the various bulb and corm crops, 50 kg/ha of P_2O_5 and 100 kg/ha of K_2O is adequate for the life of the crop, applied at planting. If the soil indices are 0 or 1, double these rates are appropriate. Magnesium should be applied at index 0 or 1 at 100 or 50 kg/ha Mg respectively. Trace element deficiencies are not found in bulb crops.

Chapter 20

FRUIT AND NURSERY STOCK

ALL THE crops covered in this chapter are perennials. For this reason fertiliser use before planting assumes more importance than for annual crops. Acidity or nutrient deficiency is not easily corrected once the crop has been planted. This is particularly important for top fruit crops which may be down for twenty-five years or longer. The crops included are top fruit, soft fruit, vines, hops, roses and other field grown hardy nursery stock. Most have an experimental basis for at least their nitrogen requirement.

GROWING SYSTEMS

Traditionally these crops have been grown in rows, under either grass or cultivation systems of management. Commonly top fruit was grassed down and the grass was grazed by sheep. Soft fruit and hops were cultivated between the rows, with mulching used to help control the weeds within the row. Today, overall grass is only common in the higher summer rainfall areas. Very few growers carry out cultivations under established fruit due to the risk of root damage. Most top fruit is grown in a herbicide strip with grass down the alleys or in overall herbicide treated soil. The latter is preferred on sites where drought is a limitation to yield. Virtually all the other crops covered in this chapter are grown under complete herbicide.

The nutrition of crops growing in soil devoid of other vegetation due to regular herbicide use presents special problems. Biological movement and cycling of nutrients is much reduced compared to soil growing a grass sward. The build-up of acidity is concentrated in the surface with the practical result that frequent small applications of lime are necessary to maintain the pH of the surface soil. If acidity is allowed to build up to more than 10 cm depth it can only be corrected very slowly by surface lime applications. For this reason, it is now recommended that soil pH is raised to 6.5 before planting fruit crops and that the pH is regularly monitored throughout the life of the crop.

The adoption of overall herbicide systems has increased yields in the absence of irrigation but has also increased other nutritional problems. Iron deficiency is more likely on susceptible sites and phosphorus uptake is sometimes reduced. The latter has implications for the storage quality of apples, discussed later in the chapter. Another specific feature of these crops is that many are grown in wide rows for management reasons. Nutrient uptake occurs predominantly from soil close to the row. Very few roots in crops like raspberries occur midway between the rows. For this reason, it is reasonable to apply nitrogen to only part of the soil surface in the orchard or plantation. If this is done, much less fertiliser is needed than if the same application rate is applied to the whole soil surface. The practicality of doing this will depend on the intensity of the growing system and the machinery available. If other nutrients are applied in this way, it will result in strips of varying soil nutrient content when the field is grubbed.

PREPLANTING FERTILISER USE

Fields intended for planting with perennial crops should be sampled to two depths, 0 to 15 cm and 15 to 30 cm. This is particularly important on land previously in fruit, hops or grass where a gradient in nutrient content and acidity will probably have built up. The 15 to 30 cm sampling is not essential on land previously ploughed regularly to 25 cm or more. Sampling should be carried out before ploughing so that if lime or fertiliser needs to be ploughed down, it can be applied first. In old herbicide strip orchards separate samples should be taken from the grass alley and the strip, especially where lime and fertiliser have previously been applied to the strip only. When sampling fields on non-calcareous soils where there is a risk of acidity, each core should be tested for pH in the field using soil indicator. Soil analysis of bulked samples will not necessarily show acid patches within the field.

Liming
Any lime requirement should be applied and cultivated in before planting. Because of the patchiness of acidity problems and the speed with which acidity can develop under herbicide management, the pH of the whole plough layer should be brought up to pH 6.5 before planting fruit or hops.

Where previously ploughed land has been sampled from 0 to 15 cm depth only, any lime requirement should be multiplied by the appropriate plough depth factor. For example, land ploughed to 25 cm should receive one and two-thirds times the 0 to 15 cm sample recommendation. Where two depths have been sampled, the two lime requirements should be added together. If the total requirement in either circumstance is above 7.5 t/ha of ground chalk or limestone, half should be ploughed down, followed by the other half cultivated in after ploughing. If the total requirement is less than 7.5 t/ha it should all be applied after ploughing and cultivated in. Appropriate pH levels

Table 20.1. Soil requirement of some nursery stock species

Acid soil (pH 4.5–5.5)	Non-calcareous soil (pH 5.5–6.5)
Andromeda	Amelanchier
Azalea (except deciduous)	Arctostaphylos
Calluna	Camellia
Cassiope	Clethra
Daboecia	Fothergilla
Enkianthus	Hydrangea (blue)
Erica (except carnea)	Lapageria
Gaultheria	Magnolia
Kalmia	Nyssa
Menziesia	Philesia
Pernettya	Stewartia (Stuartia)
Phyllodoce	
Pieris	
Rhododendron	
Vaccinium	

for fruit crops are given in Table 3.8. Nursery stock species requiring either acid or non-calcareous soil are listed in Table 20.1.

Phosphorus, Potassium and Magnesium

Before planting perennial crops it is essential that the nutrient levels in the plough layer are increased where necessary to a satisfactory level. These levels can then be maintained by surface applications for the life of the crop. If this is not performed successfully, the opportunity will have been missed and phosphorus levels in particular cannot be raised once the crop has been established.

The desired nutrient indices are 2 or above for phosphorus, 2 for potassium and 2 or above for magnesium. A soil phosphorus index of 3 may be advantageous for apples. A higher potassium index is a disadvantage for apples, so potash fertiliser should not be used at index

3 or above. If the potassium index is upper 2 or 3 a soil magnesium index of 3 is preferable to avoid induced magnesium deficiency occurring. Ideally soil nutrient levels should be sorted out a year or two ahead of planting of long-term crops. Higher indices are advantageous for hops.

The rates of nutrients given in Table 20.2 for a 15 cm depth of soil should be applied in the autumn before planting. Where previously ploughed land has been sampled to 15 cm depth only, the recommended rates should be thoroughly cultivated in before planting. Before

Table 20.2. Pre-planting fertiliser recommendations for top fruit

	N, P, K or Mg index				
	0	1	2	3	Over 3
	kg/ha				
Nitrogen (N)	Nil	Nil	Nil	—	—
Phosphate (P$_2$O$_5$)	200	100	50	50	Nil
Potash (K$_2$O)	200[a]	100	50	Nil	Nil
Magnesium (Mg)	100	75	50	Nil	Nil

(a) These rates must be applied in the autumn and well incorporated to avoid root scorch to the newly planted crop.
Source: ADAS.

planting top fruit, if the analysis shows a field to be index 0 or 1 for phosphate, potash or magnesium, the appropriate nutrient rates should be ploughed down and in addition the same rate applied and thoroughly cultivated in before planting. If plough depth is less than 20 cm, the rate ploughed down should be halved. Higher rates may be worthwhile to ensure optimum indices are achieved, particularly for hops.

Where samples have been taken from 0 to 15 cm depth and 15 to 30 cm, the appropriate nutrient rates should be ploughed down before top fruit is planted if the 15 to 30 cm sample is index 0 or 1 for phosphate, potash or magnesium. After ploughing, the rate based on the 0 to 15 cm sample should be applied and thoroughly cultivated in before planting. If plough depth is less than 20 cm the rate ploughed down should be halved. Where it is not possible to plough fertiliser down, the application should be limited to the amount recommended for one sampling depth only. It is not necessary to plough fertiliser down before planting soft fruit or nursery stock. Nitrogen is not required before planting.

FERTILISERS FOR ESTABLISHED CROPS

For all established crops sampling depth should be to 15 cm. Where soil has been undisturbed for a number of years, sampling depth must be as accurate as possible, and must include the top 5 cm layer. A tubular corer is best for this purpose. Orchards in overall grass or overall herbicide management should be sampled within the spread of the tree branches. In orchards with herbicide strip management sampling should be restricted to the strip, excluding the grass area. Samples from soft fruit plantations, hops and nursery stock should be taken at random from within the area of rooting.

Nitrogen

Modest annual applications of nitrogen fertiliser are adequate to maintain the nitrogen balance of established perennial crops grown on overall herbicide treated land. Recommendations are based on both experiments and experience. Higher rates are generally needed for crops that need to produce a lot of new growth annually such as hops and raspberries, compared to those that produce much less such as apples and strawberries. The amount of nutrients removed annually in crop and prunings varies considerably.

Where excess nitrogen is given, tree and bush fruits are liable to be over vegetative with large, dark green leaves. Apple quality, especially taste, firmness and storage qualities, may be adversely affected. High nitrogen reduces the amount of red colour and intensifies the green colour of apples. This effect is detrimental to crop appearance and value in red-coloured varieties, but can be beneficial in culinary varieties such as Bramley.

Very high levels of nitrogen manuring can reduce the alpha-acid content of hop cones, although up to a certain point it will produce more alpha-acid per hectare because the crop itself is greater. Where progressive Verticillium wilt is present high rates of nitrogen will make the hops more susceptible to attack from this disease. On farms where there is a risk of wilt, it will be necessary to reduce applications to below those normally recommended for maximum yield.

Where a medium or high grade shoddy (10 to 12 per cent N) has been applied regularly, the total nitrogen application (shoddy plus fertiliser N) should be adjusted, assuming two-thirds of the nitrogen content of the shoddy to be available in the year of application. Applications of straw reduce the amount of nitrogen available to the crop because nitrogen is used by micro-organisms in the decomposition of the straw. This should be allowed for by increasing the amount of

Table 20.3. Nitrogen recommendations for established perennial crops

Crop	Nitrogen kg/ha per year
Desert apples – overall herbicide	40
– herbicide strip	60
Culinary and cider apples – overall herbicide	100
– herbicide strip	120
– overall grass	180
Cherries, pears and plums – overall herbicide	120
– herbicide strip	150
Cane fruits, redcurrants, gooseberries	100
Blackcurrants	140
Strawberries	Nil
Vines	40
Hops	225
Roses – rootstock (1st year)	100
– after budding (2nd year)	50
Slow growing nursery stock – Ericas, Azaleas, Rhododendrons	50
Medium growing nursery stock – Berberis, Ilex, Senecio	100
Fast growing nursery stock – Pyracantha, Hydrangea, Leyland's Cypress	150

Source: ADAS.

nitrogen applied by 10 kg for every tonne of straw used, particularly if the straw is rotavated into the soil.

The recommendations in Table 20.3 are a guide to use on perennial crops. For top fruit, no nitrogen is required for the first three years after planting. For other crops apply nitrogen in the first year. Nitrogen for perennial crops should usually be applied in March or April. There is some evidence of a benefit from splitting the application for hops, but otherwise a single application is satisfactory. Autumn applications are sometimes recommended for fruit crops but these are difficult to justify. Crop uptake of autumn-applied nitrogen will invariably be very poor.

Phosphorus
Assuming a soil index of 2 or 3 established crops need a maintenance application of 20 kg/ha/year P_2O_5 for top fruit and 40 kg/ha/year P_2O_5 for soft fruit. Timing of application does not matter. If more convenient, the requirement for three years can be given in one application. At higher indices, phosphate fertiliser is not needed except for hops, which always attract a higher recommendation.

Maintenance applications are only justified for nursery stock if the crop is down for more than three years, as long as adequate application was applied before planting.

Potassium

Appropriate potash fertiliser depends on species and soil type. The light and medium soil textures on which most of these crops are concentrated need regular potassium to ensure that the soil level is maintained in index 2. At the upper end of index 2, application may be reduced if apple quality problems are encountered. For top fruit, blackberries, strawberries and vines, 80 kg/ha/year K_2O is adequate for maintenance. For blackcurrants and other fruit crops, apply 120 kg/ha/year K_2O. Again more generous amounts are normally recommended for hops. For nursery stock on sandy soils apply 50 kg/ha/year K_2O. On heavier soils this can be omitted unless the crop is down for more than three years.

Raspberries, redcurrants and gooseberries may be damaged by high spring applications of chloride-containing fertilisers. Otherwise timing of potash application is not crucial.

Magnesium

At index 2 apply a maintenance application of 30 kg/ha per year of magnesium. Higher rates should be applied at lower indices which were not built up before planting. Magnesium can often be supplied cheaply by using magnesian limestone when the soil needs liming. When liming is not required, kieserite or calcined magnesite should be used. Where magnesium deficiency has been diagnosed, foliar sprays of agricultural magnesium sulphate (Epsom salts) may be used to give more rapid control than soil dressings of magnesium fertilisers. Depending on the degree of the deficiency, two to five applications of 20 kg/ha magnesium sulphate in 1,000 litres may be necessary, applied at fourteen-day intervals.

Trace Elements

Trace element deficiencies may occur in fruit and in nursery stock. These deficiencies can often be identified by visual diagnosis but should be checked by leaf and soil analysis, except in the case of iron deficiency which cannot easily be confirmed by analysis.

Boron deficiency in fruit crops is rare. There is some recent evidence suggesting a possible association between boron deficiency and some forms of apple fruit cracking. Storage life of apples may be reduced if unnecessary boron is applied. Where deficiency is confirmed, the best treatment is foliar application of Solubor at the rate of 2 kg/ha in 1,000 litres plus wetter. Three sprays starting at petal fall and repeated at two- to three-week intervals are recommended. The deficiency can also be corrected by a soil application of 20 kg/ha borax or 10 kg/ha Solubor in the spring.

Copper deficiency in pears has been diagnosed on occasions in orchards on sandy soils. It can be corrected by applying a foliar spray of copper oxychloride or cuprous oxide fungicide in late May at the rate of 2 kg/ha in 1,000 litres.

Iron deficiency can best be corrected in top fruit by applying 60 g of iron-EDDHA chelate per tree to the soil within the area of branch spread in early February and subsequent repeated annual applications of 30 g per tree. Iron-EDTA can be used as a foliar spray at 1 kg/ha in 1,000 litres plus wetter. Four or five sprays starting at petal fall and repeated at fourteen-day intervals may be necessary. Iron deficiency is common in soft fruit crops and several species of nursery stock when grown on calcareous soils. It is best corrected by an annual application of 8 g per square metre of iron-EDDHA chelate applied to the rooting zone of the crop in early February. Alternatively, iron-EDTA can be used as a foliar spray at 1 kg/ha in 1,000 litres plus a wetter. Where the deficiency is severe it may be necessary to repeat the spray treatment at fourteen-day intervals.

Manganese deficiency is best controlled by a foliar spray of manganese sulphate at 4 kg/ha in 1,000 litres plus a wetter. This concentration should never be exceeded. One spray (at petal fall in top fruit) is usually sufficient but if the symptoms persist a further spray can be given.

Zinc deficiency has very occasionally been found to reduce growth and cropping of apple trees and forest nursery stock on sandy soils. This can be corrected in apples by a spray of zinc sulphate at green cluster or petal fall at 1 kg/ha in 1,000 litres plus wetter.

LEAF ANALYSIS FOR FRUIT CROPS

Leaf analysis is a very useful technique for the diagnosis of nutritional disorders. Samples should be taken in a similar manner from both good and poor areas of growth so that the results can be compared. In addition, the development of satisfactory ranges of leaf nutrient concentration for optimum growth and cropping has provided a method of assessing the nutritional status of crops. Satisfactory nutrient levels are given in Table 20.4. Where results are to be compared to the standards, it is essential that a representative sample is taken at the correct time.

Table 20.4. Leaf analysis—satisfactory nutrient ranges for fruit crops

Crop	Leaf sampling position[a]	Nitrogen (N)	Phosphorus (P)	Potassium (K)	Magnesium (Mg)
		Percentage of nutrient in dry matter			
Apples					
Cox	1	2.6–2.8	0.20–0.25	1.2–1.6	0.20–0.25
Bramley	1	2.4–2.8	0.18–0.23	1.2–1.6	0.20–0.30
Cherries	1	2.4–2.8	0.20–0.25	1.5–2.0	0.20–0.25
Pears					
Comice	1	1.8–2.1	0.15–0.20	1.2–1.6	0.20–0.25
Conference	1	2.1–2.6	0.15–0.20	1.2–1.6	0.20–0.25
Plums	1	2.0–2.6	0.15–0.20	1.5–2.0	0.20–0.25
Blackcurrants	2	2.8–3.0	0.25–0.35	1.5–2.0	0.15–0.25
Raspberries	3	2.4–2.8	0.20–0.25	1.5–2.0	0.30–0.35
Strawberries	4	2.6–3.0	0.25–0.30	1.5–2.0	0.15–0.20

(a) Leaf sampling position:
  1. Mid-third extension growth, sampled mid-August.
  2. Fully expanded leaves extension growth, sampled prior to harvest.
  3. Fully expanded leaves non-fruiting canes, sampled at fruit ripening.
  4. Lamina of recently matured leaves, sampled at fruiting stage.
Source: ADAS.

There are seasonal and other factors which influence leaf nutrient levels, therefore leaf analysis must be interpreted with caution. Leaf nutrient levels can also vary between varieties. Where there is sufficient information the standard ranges take account of varietal differences.

Leaf analysis can be used to provide a more complete guide to the adequacy of the orchard fertiliser programme than can be obtained from soil analysis alone. Where leaf nutrient levels are below the satisfactory range an increase in fertiliser use can be considered. However, before making a change, the cause should be further investigated to ensure that other factors such as soil compaction or disease are not involved.

Where the leaf nutrient level is consistently above the satisfactory range for several years there may be justification for a reduction in fertiliser use. High levels of nitrogen and potassium can have adverse effects on apple storage quality and application rates can often be reduced to advantage.

Leaf manganese is commonly analysed. Below 20 mg/kg is deficient while a level over 100 mg/kg is above normal and indicates a need to check soil pH in the orchard. High manganese levels can also result from the use of foliar feeds or fungicides containing manganese.

APPLE FRUIT ANALYSIS

Analysis of fruit sampled within two weeks of picking has proved a useful indicator of the risk of some physiological disorders in stored apples. Results can also be used to rank orchards for potential storage quality. For Cox and Bramley, the standards in Table 20.5 have been developed for fruit suitable for long-term storage.

As for leaf analysis, if fruit analysis produces consistently low or high figures for a particular nutrient over two or three years, modification of fertiliser application should be considered. The most likely change will be a reduction in nitrogen or potassium use. Fruit analysis may also show deficiencies of calcium or phosphorus which can reduce fruit storage quality. These deficiencies can be corrected by foliar sprays of calcium and phosphorus or by post-harvest calcium treatments.

Table 20.5. Apple fruit—satisfactory nutrient levels for storage

	Nitrogen (N)	Phosphorus (P) (minimum)	Potassium (K)	Magnesium (Mg)	Calcium (Ca) (minimum)	
			mg/100 g fresh weight			
Cox	50–70	11.0	130–160	5.0	4.5[a]	5.0[b]
Bramley	50–60	9.0	105–115	5.0	4.5[a]	5.0[b]

(a) For controlled atmosphere storage (Cox in 2 per cent oxygen until late March; Bramley in 8 per cent carbon dioxide until May)
(b) For storage in air at recommended temperatures (Cox until mid December; Bramley until late January).
Source: ADAS.

APPENDICES

APPENDIX I

Metric to Imperial Conversions

1 kg/ha	= 0.8 units/acre
	= 0.9 lb/acre
1 t/ha	= 0.4 ton/acre
	= 8 cwt/acre
1 l/ha	= 0.09 gall/acre
1 kg/t	= 2 units/ton
1 kg/m³	= 9 units/1,000 galls
1 kg/1,000 litres	= 9 units/1,000 galls

APPENDIX II

Chemical Nomenclature

Elements

Aluminium	Al
Arsenic	As
Boron	B
Cadmium	Cd
Calcium	Ca
Carbon	C
Chlorine	Cl
Chromium	Cr
Cobalt	Co
Copper	Cu
Fluorine	F
Hydrogen	H
Iodine	I
Iron	Fe
Lead	Pb
Magnesium	Mg
Manganese	Mn
Mercury	Hg
Molybdenum	Mo
Nickel	Ni
Nitrogen	N
Oxygen	O
Phosphorus	P
Potassium	K
Selenium	Se
Silicon	Si
Sodium	Na
Sulphur	S
Zinc	Zn

Ions

Ammonium	NH_4^+
Bicarbonate	HCO_3^-
Chloride	Cl^-
Ferric	Fe^{3+}
Ferrous	Fe^{2+}
Hydroxide	OH^-
Iodate	IO_3^-
Iodide	I^-
Molybdate	MoO_4^{2-}
Nitrate	NO_3^-
Phosphate	HPO_4^{2-}
	$H_2PO_4^-$
Selenate	SeO_4^{2-}
Sulphate	SO_4^{2-}
Sulphide	S^-

Compounds

Ammonia	NH_3
Boric acid	$B(OH)_3$
Calcium carbonate	$CaCO_3$
Calcium hydroxide	$Ca(OH)_2$
Calcium oxide	CaO
Calcium silicate	$CaSiO_3$
Calcium sulphate	$CaSO_4$
Carbon dioxide	CO_2
Hydrogen sulphide	H_2S
Magnesium oxide	MgO
Nitrous oxide	N_2O
Phosphorus pentoxide	P_2O_5
Potassium oxide	K_2O
Silicon dioxide	SiO_2
Sulphur dioxide	SO_2
Urea	$CO(NH_2)_2$
Water	H_2O

INDEX

FARMING PRESS BOOKS

Below is a sample of the wide range of agricultural and veterinary books published by Farming Press. For more information or a free illustrated book list please contact:

Books Department, Farming Press Ltd, Wharfedale Road, Ipswich IP1 4LG, Suffolk, Great Britain.

Cereal Pests and Diseases
R. Gair, J. E. E. Jenkins and E. Lister

An outstanding guide to the recognition and control of cereal pests and diseases by two of Britain's foremost plant pathologists and a leading entomologist.

Farm Woodland Management
John Blyth, Julian Evans, William E. S. Mutch and Caroline Sidwell

A compendium for farmers in which all aspects of trees on the farm are considered: planting bare ground as well as dealing with existing woodland; financial aspects; environmental considerations.

Direct Drilling and Reduced Cultivations
H. Allen

Outlines the technical advances which have been made and summarises the relevent research findings on the effects of direct drilling. Practical techniques are considered in detail for cereals, oilseed rape, grass, forage crops, sugarbeet and some overseas crops.

Soil Management
D. B. Davies, D. J. Eagle and B. Finney

Two soil scientists and a senior mechanisation officer with ADAS go into all aspects of the soil, plant nutrition, farm implements and their effects on the soil, crop performance, land drainage and cultivation system.

Potato Mechanisation and Storage
C. F. H. Bishop and W. F. Maunder

A comprehensive look at the latest techniques and equipment available for potato growers.

Oilseed Rape
J. T. Ward, W. D. Basford, J. H. Hawkins and J. M. Holliday

Contains up-to-date information on all aspects of oilseed rape growth, nutrition, pest control and marketing.

Intensive Sheep Management
Henry Fell

An instructive practical account of sheep farming based on the experience of a leading farmer and breeder.

Farm Machinery
Brian Bell

Gives a sound introduction to a wide range of tractors and farm equipment.

The Farm Office: How to Cope with Paperwork
Michael Hosken

From preparing annual accounts to choosing a filing cabinet, this book covers virtually every aspect of day-to-day farm office work and is particularly useful for the training of farm secretaries.

Straw for Fuel, Feed and Fertiliser
A. R. Staniforth

Provides an excellent practical summary of the profitable disposal and utilisation of straw for arable and livestock farmers.

Farming Press also publish three monthly magazines: *Arable Farming*, *Pig Farming* and *Dairy Farmer*. For a specimen copy of any of these magazines please contact Farming Press at the address above.